D1175977

THE BIOLOGY
of the
HETEROPTERA

THE BIOLOGY

OF THE

HETEROPTERA

by

N. C. E. MILLER
formerly
Government Entomologist, Department of Agriculture
Straits Settlements and Federated Malay States
and
Senior Entomologist
Commonwealth Institute of Entomology
London

Second (Revised) Edition

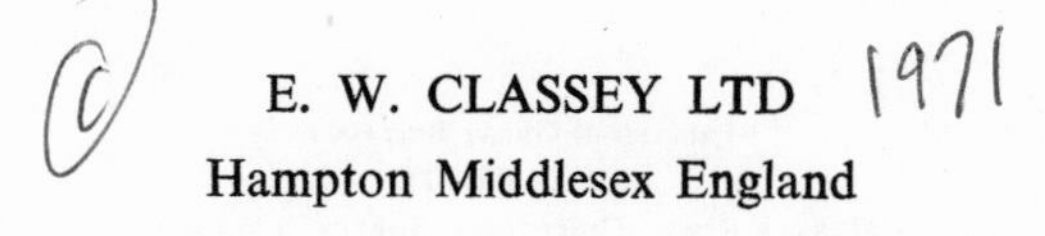

E. W. CLASSEY LTD
Hampton Middlesex England

N. C. E. Miller

THE BIOLOGY OF THE HETEROPTERA

Second (Revised) Edition

ISBN 0 900848 45 6

Printed in Great Britain by
BIDDLES LTD
Martyr Road Guildford Surrey England

CONTENTS

LIST OF PLATES

LIST OF FIGURES

INTRODUCTION TO FIRST EDITION

The Hemiptera are divided into two main groups, the Heteroptera and the Homoptera. This division is based on the structure of the mesothoracic or forewings (hemelytra) the basal part of which (corium) is almost entirely coriaceous and the apical part (membrane) membranous.

However, many types of hemelytra are represented, including some in which the corium exhibits considerable diversity in area and structure for, example in the subfamilies Holoptilinæ and Tribelocephalinæ (Reduviidæ) and in the family Enicocephalidæ.

In the Homoptera the forewings are uniform in structure.

In the classification of the Heteroptera several attempts have been made to find a satisfactory natural arrangement. They have been divided by Dufour into three series, the Geocorisæ (land bugs), Hydrocorisæ (aquatic bugs) and Amphibiocorisæ (bugs living on the surface of the water).

In the first-mentioned group fall all terrestrial bugs and into the second group those living on or in the water. Most of these, however, do leave one area of water to fly to another from time to time.

Aquatic Heteroptera are confined mainly to fresh water, but to quote exceptions, *Halobates* species occur on the sea and the curious *Aepophilus* (Aepophilidæ) lives in rock crevices in the inter-tidal zone of the seashore.

Associated with the Hydrocorisæ are two families of shore bugs (Gelastocoridæ and Ochteridæ), while the largely littoral Saldidæ are probably more correctly associated with the Hydrocorisæ.

Up to the present time, some 25,000 species of Heteroptera are known. While most of them are distributed in the tropics, some families are better represented in the Holarctic Region and some are extremely restricted in range. Fossil remains of Heteroptera have been found in the Carboniferous.

In size the Heteroptera range from about 2 to 100 millimetres in length; among the giants of the insect world are the aquatic genera *Lethocerus* Mayr 1852, *Hydrocyrius* Spinola 1850 and *Belostoma* Latreille 1807 (Belostomatidæ), while the smallest representatives are found in the Dipsocoridæ, Schizopteridæ and Helotrephidæ.

Some species are cosmopolitan having attained their wide distribution in most instances through the agency of man. Among these are *Nezara viridula* Linnæus 1758 (Pentatomidæ), *Liorhyssus hyalinus* (Fabricius) 1794 (Coreidæ), *Nabis capsiformis* Germar 1837 (Nabidæ), *Trigonotylus brevipes* Jakowleff 1880, *Lygus apicalis*

Fieber 1861 (Miridæ) and *Amphibolus venator* Klug 1830 (Reduviidæ). There are also the well-known bed-bug *Cimex lectularius* Linnæus 1758 (Cimicidæ) which is to be found in almost every land, and the aquatic *Halobates micans* Eschscholtz 1822 (Gerridæ) which is known from several oceans.

Terrestrial species are found on plants on various parts of which they feed. Some are carnivorous and feed on other Arthropods; others feed on mammalian and avian blood; some species are termitocolous or myrmecophilous.

There are no true cavernicolous Heteroptera. Several species recorded from caves have been found usually not very far from the entrance and not often in the more remote parts where it is quite dark. No species has yet been recorded exhibiting characters which indicate that it has become adapted to a life in obscurity, with the possible exception of an aberrant member of the Tribelocephalinæ, *Xenocaucus* China and Usinger 1949, which has no compound eyes, from which fact it has been assumed that this genus lives in the soil and therefore in complete absence of light. Species feeding on mammalian and avian blood belong to the subfamily Triatominæ. Some of these are eminently important from the medical point of view, since they are vectors of human trypanosomiasis.

The Heteroptera are provided with salivary glands situated mainly in the thorax, but sometimes these glands extend partly into the head or abdomen. In phytophagous species the saliva is injected into the tissues of the host-plant when the bug pierces them with its mouth-parts. It acts as an enzyme which breaks down the starch in the sap.

The effect of the saliva on plant tissues is mostly of one type. The area around the site of puncture changes colour fairly rapidly from brown to black. Following this, the affected part soon wilts and in the event of the damage being extensive the entire plant may perish. When leaves are pierced it is not uncommon for the area surrounding the site of puncture to dry up and split.

In view of the damage which quite a few Heteroptera are capable of inflicting on plants, it is obvious that they play an important rôle economically. The saliva of the predaceous species acts both as an anticoagulin and also lethally; its action on man provokes intense pain.

Certain predaceous Heteroptera may play a role, albeit a relatively minor one in reducing the numbers of an insect pest, but since they are to a very great extent, polyphagous, the effect of their depredations is rarely, if ever, obvious. A predator also, is soon satisfied and may not seek another victim for some hours, or even days, after it has consumed it.

Glands variously known as ‘scent’, ‘odoriferous’, ‘stink’ or 'repugnatorial' are present in most Heteroptera both in the adults and in the immature stages. Some Heteroptera secrete a glutinous substance through glandular setæ and some, a white wax-like substance which possibly has some connection with ecdysis.

A glutinous substance is produced by some species at the time of oviposition. This serves either to fix the ovum to the substratum or possibly to provide protection against hymenopterous parasites. It may also act as a protective covering to check loss of moisture.

Development in the Heteroptera is of the Hemimetabolous (Paurometabolous) type. In this the adult stage is preceded by stages which resemble it to some extent. During the course of development the external changes which take place (apart from increase in size) affect mainly the antennæ, wings and legs. These changes are dealt with in the chapter on Development.

Some difference of opinion exists regarding the use of the term 'larva' to denote the developmental stages between the ovum and the adult. For the apterygota and the exopterygota certain authors have adopted the name 'nymph' which they consider to be applicable to all the stages of post-embryonal development. Other authors, however, restrict the designation 'nymph' to the ultimate and penultimate instars in which the wing rudiments become visible.

Berlese has suggested the names 'prosopon' or 'prosopide' for the developmental stages, but on account of the fact, it would seem, that they are not easily pronounced, they have not been widely accepted. In an attempt to settle the matter, Grandi coined a new term 'neanide' which has been found acceptable and has subsequently been used by several authors. I would suggest that all hemipterists should employ it. The term 'nymph' I consider to be singularly inappropriate.

Although the volume of work on the Heteroptera has attained considerable proportions chiefly from the systematic aspect, our knowledge of the biology of the sub-Order is extremely limited. This, I believe, is mainly because attention has been chiefly confined to the collection of specimens and not to recording at the time, facts concerning the habits and habitats of the species concerned.

In this book I have attempted to collate as fully as possible the biological information which is scattered in the voluminous literature. Added to this are personal observations made in the field in various parts of eastern and central Africa and in Malaysia.

It has not been practicable, obviously, to quote all the literature both on the systematics and biology of the Heteroptera, but it is hoped that the references given will provide a basis for further studies.

With regard to the names of the families and subfamilies, it is pointed out that they are formed from the oldest group name based on a valid genus. Each family is dealt with separately to facilitate reference and to avoid repetition.

For other papers on the developmental stages of Heteroptera, the *Bulletin of Entomological Research, The Review of Applied Entomology, The Journal of Economic Entomology* and the publications of agricultural organisations in the Commonwealth and Empire should be consulted.

In a recent work, Cobben has dealt very extensively with the structure and biology of the ova of the Heteroptera. It is not unreasonable to suppose that this contribution may help to throw more light on the very obscure subject of their evolution.

In the preparation of this book I have had valuable and freely given advice from my colleague, the eminent hemipterist Dr. W. E. China, Deputy Keeper of the Department of Entomology, British Museum (Natural History), London. I have also had access to the very extensive collections, card-indices and library of the British Museum. For this very desirable privilege I am greatly indebted to the Trustees.

Finally, I express my thanks to the late Sir Harold Tempany, C.M.G., C.B.E., formerly Director of Agriculture, Straits Settlements and Federated Malay States, and later Agricultural Adviser to the Colonial Office, for his much appreciated support; and to Dr. W. J. Hall, M.C., C.M.G., Director of the Commonwealth Institute of Entomology, London, for permission to carry out the work.

The coloured plate and all the line figures have been prepared by myself, except in those cases in which the authorship is acknowledged. For the photographic plates I have to thank the Photographic Department of the British Museum.

References

Amyot and Serville 1843; Berlese 1914; Butler 1923; Carayon 1962; China 1933, 1943, 1955b; China and Miller 1955; Cobben 1968; Costa Lima 1940; Distant 1904; Douglas and Scott 1865; Dufour 1833; Ekblom 1926; Evans 1948; Grandi 1951; Handlirsch 1908; Hemming 1953; Horvath 1911, 1912; Hungerford 1919; Imms 1934; Jeannel 1919; Kirkaldy 1906; Kalshoven 1950; Kirkaldy 1899; Maxwell-Lefroy 1909; Miller 1931b, 1934, 1953; Myers 1926; Oshanin 1912; Poisson 1924, 1935, 1951; Readio 1927; Saunders 1892; Southall 1730; Spinola 1837; Thomas 1954; Usinger 1946; Villiers 1952; Weber 1930; Wigglesworth 1939; Reuter 1912.

INTRODUCTION TO SECOND EDITION

Since the publication in 1956 of this work, a considerable amount of new information, as might be expected, has been published. It has therefore been considered that the appropriate time has arrived to add it to that contained in the First Edition.

The opportunity has been taken to correct errors, some of them regrettably avoidable, and also to refer to works which had been overlooked.

The information contained in works not previously referred to but now included in the list of references, embraces studies on the evolution of water-bugs, on the salivary glands, on the effect of saliva on plant growth and on wheat grains. Methods of respiration, the mechanism of ecdysis, the function of egg-bursters and the rôle of Heteroptera as pollinators are also innovations.

The list of families and subfamilies has been brought up to date and descriptions of new families are included, as well as figures, some of which for the purpose of portraying more clearly certain species shown on the plates.

I am greatly indebted to my many colleagues for drawing my attention to matter which should have been included in the First Edition had it not been overlooked.

I also wish to express my grateful thanks to Mrs. Gillian Black, British Museum (Natural History), London, for assistance in connection with nomenclature matters.

References

China 1955; Elson 1937; Kiritshenko 1949; Massee 1956; Myers 1926; Nuorteva 1956a, 1956b, 1958; Parshley 1917; Schumacher 1917; Thorpe 1950; Thorpe and Crisp 1947.

PART 1

GENERAL ACCOUNT OF THE HETEROPTERA

PART 1

Chapter 1

Family and Subfamily Names of Heteroptera

Order **Hemiptera**

Suborder *Heteroptera*

Family Plataspidæ Dallas 1851, *Cat. Hemipt. Brit. Mus.* **1,** 61.
Syn. Arthropteridæ Fieber 1860, *Europ. Hemipt.* 27.
Syn. Coptosominæ Kirkaldy 1909, *Cat. Hemipt.* **1,** Cimicidæ, 36.
Syn. Coptosomatidæ Reuter 1912, *Öfv. Finska Vet. Soc. Förh.* **54a,** No. 6, 45.
Syn. Brachyplatidæ Leston 1952, *Ann. Mag. nat. Hist.* (12), **5,** 512.
Family Lestoniidæ (China) 1955, *Ann. Mag. nat. Hist.* (12), **8,** 210 (Lestoniinæ).
Family Cydnidæ (Billberg) 1820, *Enum. Ins. Billb.* 70 (Cydnides).
Subfamily 1. Thyreocorinæ (Amyot and Serville) 1843, *Hist. nat. Hémipt.*, xviii, 60 (Thyreocorides).
Syn. Corimelæninæ Uhler 1872, *Report U.S. Geol. Survey* (1871), **4,** 471.
Syn. Thyreocorinæ Van Duzee 1907, *Bull. Buffalo Soc. nat. Sci.* **8,** pt. 5, 5.
Subfamily 2. Cydninæ (Amyot and Serville) 1843, *Hist. nat. Hémipt.* xx, 87 (Cydnides).
Syn. Cydninæ Dallas 1851, *Cat. Hemipt. Brit. Mus.* **1,** 109 (*partim*).
3. Sehirinæ (Amyot and Serville) 1843, *Hist. nat, Hémipt.* xxi, 96 (Sehirides). Sehirinæ (Stål) 1864. *Hemipt. Afric.* **1,** 18 (Sehirida).
Syn. Cydninæ Dallas, *Cat. Hemipt. Brit. Mus.* **1,** 109 (*partim*).
Family Pentatomidæ (Leach) 1815, *Brewster's Edinb. Encyc.* **9,** 121 (Pentatomides).
Syn. Cimicidæ Kirkaldy 1909, *Cat. Hemipt.* **1,** 1.
Subfamily 1. Asopinæ (Amyot and Serville) 1843, *Hist. nat. Hémipt.* xix, 77 (Asopides).
Syn. Asopinæ (Spinola) 1850, *Tav. Sin. Hem. ex Mem. Math. Fis. Soc. Ital. Sci. Modena* **25,** 1; 69 (1852) (Asopoideæ).

Syn. Amyotinæ Schouteden 1906, *Wytsman Gen. Ins.* **52,** 2.
Syn. Amyotinæ Leston 1953, *Ent. Gaz.* **4,** 19.
Syn. Arminæ Bergroth 1908, *Mem. Soc. ent. belg.* **15,** 180.
Syn. Tahitocorinæ (Yang) 1935, *Ann. Mag. nat. Hist.* (10), **16,** 480 (Tahitocoridæ).
2. Tessaratominæ (Stål) 1865, *Hemipt. Afric.* **1,** 33 (Tessaratomida).
3. Eumenotinæ Esaki 1922, *Ins. Insc. Mens.* **1,** 196 (under Aradidæ); 1930 *Ann. Mag. nat. Hist.* (10) **5,** 630 (under Pentatomidæ).
4. Cyrtocorinæ Distant 1880, *Biol. centr. Amer. Rhynchota, Het.* **1,** 43.
5. Dinidorinæ (Stål) 1870, *Enum. Hemipt.* 79 (Dinidorida).
6. Phyllocephalinæ (Amyot and Serville) 1843, *Hist. nat. Hémipt.* xxix, 174 (Phyllocephalides).
Syn. Phyllocephalinæ (Dallas) 1851, *Cat. Hemipt. Brit. Mus.* **1,** 350 (Phyllocephalidæ).
7. Pentatominæ (Amyot and Serville) 1843, *Hist. nat. Hémipt.* xxiv, 124 (Pentatomides).
Syn. Pentatominæ Stål 1864, *Hemipt. Afric.* **1,** 32, 76.
Syn. Halydidæ Dallas 1851, *Cat. Hemipt. Brit. Mus.* **1,** 150.
Syn. Sciocoridæ Dallas 1851, *Cat. Hemipt. Brit. Mus.*, **1,** 130.
Syn. Sciocorides Amyot and Serville 1843, *Hist. nat. Hémipt.* 118.
Syn. Macropeltidæ Fieber 1860, *Europ. Hemipt.* 26, 327.
Syn. Discocephalidæ Fieber 1860, *Europ. Hemipt.* 26, 326.
Syn. Aeliidæ Douglas and Scott 1865, *Brit. Hemipt. Heteroptera*, 14.
Syn. Rhaphigastridæ (Amyot and Serville) 1843, *Hist. nat. Hémipt.* xxv, 141 (Rhaphigastrides).
8. Scutellerinæ (Leach) 1815, *Brewster's Edinb. Encyc.* **9,** 121 (Scutellerida).
Syn. Pachycorinæ (Amyot and Serville) 1843, *Hist. nat. Hémipt.* xvi, 34 (Pachycorides).
Syn. Tetyrinæ (Amyot and Serville) 1843, *Hist. nat. Hémipt.* xvii, 45 (Tetyrides).
Syn. Eurygastrinæ (Amyot and Serville) 1843, *Hist. nat. Hémipt.* xviii, 51 (Eurygastrides).
Syn. Odontoscelidæ Douglas and Scott 1865, *Brit. Hemipt. Heteroptera*, 13.
Syn. Eurygastridæ Douglas and Scott 1865, *Brit. Hemipt. Heteroptera*, 13.

9. Podopinæ (Amyot and Serville) 1843, *Hist. nat. Hémipt.* xviii, 56 (Podopides).
Syn. Oxynotidæ (Amyot and Serville) 1843, *Hist. nat. Hémipt.* xviii, 58 (Oxynotides).
Syn. Graphosomatinæ Jakowleff 1884, *Hor. Soc. ent. Ross.* **18,** 204.
10. Serbaninæ Leston 1953, *Rev. Brasil Biol.* **13,** 137.
11. Acanthosomatinæ (Stål) 1864, *Hemipt. Afric.* **1,** 33, 219 (Acanthosomida).
12. Canopinæ (Amyot and Serville) 1843, *Hist. nat. Hémipt.* xix, 70 (Canopides).
Syn. Canopinæ Horváth 1919, *Ann. Mus. nat. Hung.* **17,** 205.
13. Megaridinæ McAtee and Malloch 1928, *Proc. U.S. nat. Mus.* **72,** 1.

Family Aphylidæ (Bergroth) 1906, *Zool. Anz.* **29,** 646 (Aphylinæ).
Syn. Aphylidæ Reuter 1912, *Öfv. Finska Vet. Soc. Förh.* **54a,** No. 6, 46.

Family Urostylidæ Dallas 1851, *Cat. Hemipt. Brit. Mus.* **1,** 313.
Syn. Urolabidæ (Stål) 1876, *Svensk. Vet.-Ak. Handl.* **14,** 4; 115 (Urolabida).
Syn. Urolabididæ Reuter 1912, *Öfv. Finska Vet. Soc. Förh.* **54,** No. 6; 37.

Subfamily 1. Urostylinæ Dallas 1851, *Cat. Hemipt. Brit. Mus.* **1,** 313.
2. Saileriolinæ China and Slater 1956, *Pacific Sci.* **10,** 412.

Family Phloeidæ (Amyot and Serville) 1843, *Hist. nat. Hémipt.* xxiv, 115 (Phloeides).
Syn. Phlœidæ Dallas 1851, *Cat. Hemipt. Brit. Mus.* **1,** 419.

Family Coreidæ Leach 1815, *Brewster's Edinb. Encyc.* **9,** 121.
Syn. Lygæidæ Kirkaldy 1899, *Entomologist* **32,** 220.

Subfamily 1. Meropachydinæ (Stål) 1867, *Öfv. Vet.-Ak. Förh.* **24,** 535–536 (Meropachydida).
Syn. Merocorinæ (Stål) 1870, *Enum. Hemipt.* **1,** 125 (Merocorina).
2. Coreinæ (Stål) 1867, *Öfv.-Vet.-Ak. Förh.* **24,** 535–543 (Coreida).
Syn. Centroscelinæ Kirkaldy 1899, *Entomologist,* **32,** 220.
Syn. Anisoscelidæ Dallas 1852, *Cat. Hemipt. Brit. Mus.* 449.
3. Pseudophlœinæ (Stål), *Öfv. Vet.-Ak. Förh.* **24,** 535 (Pseudophlœida).
Syn. Arenocorinæ Bergroth 1913, *Mem. Soc. ent. belg.* **22,** 135.

4. Agriopocorinæ Miller 1953, *Proc. Linn. Soc. N.S Wales*, **78,** pts. 5 and 6; 233.
5. Rhopalinæ (Amyot and Serville) 1843, *Hist. nat. Hémipt.* xxxiii, 243 (Rhopalides).

Syn. Rhopalidæ Dallas 1852, *Cat. Hemipt. Brit. Mus.* 520.
Syn. Corizidæ Douglas and Scott 1865, *Brit. Hemipt. Heteroptera* 17.
Syn. Chorosomidæ Douglas and Scott 1865, *Brit. Hemipt. Heteroptera* 17.
Syn. Corizinæ Mayr 1866, *Reise Freg. Novara, Zool. Hemipt.* **2,** 121.

6. Alydinæ (Amyot and Serville) 1843, *Hist. nat. Hémipt.* xxxiv, 221 (Alydides).

Syn. Alydinæ (Dallas) 1852, *Cat. Hemipt. Brit. Mus.* **2,** 467 (Alydidæ).
Syn. Coriscidæ Stichel 1925, *Illus. Bestimmungstabellen Deutsch. Wanz.* 45.

Family Stenocephalidæ Dallas 1852, *Cat. Hemipt. Brit. Mus.* **2,** 480.
Syn. Dicranocephalidæ (Scudder) 1957, *Proc. R. ent. Soc. London* (A) **32,** 147 (Dicranocephalini).

Family Hyocephalidæ Bergroth 1906, *Zool. Anz.* **29,** 649.

Family Lygæidæ (Schilling) 1829, *Beitr. z. Ent.* **1,** 37 (Lygæides).
Syn. Myodochidæ Kirkaldy 1899, *Entomologist* **32,** 220.
Syn. Geocoridæ Kirkaldy 1902, *Journ. Bombay Nat. Hist. Soc.* **14,** 306.
Syn. Pyrrhocoridæ Kirkaldy 1904, *Entomologist* **37,** 280.

Subfamily 1. Megalonotinæ Slater 1957, *Bull. Brooklyn ent. Soc.* **52,** 2; 35.
Syn. Rhyparochrominæ (Amyot and Serville) 1843, *Hist. nat. Hémipt.* xxxvi, 251 (Rhyparochromides).
Syn. Rhyparochrominæ (Stål) 1862, *Öfv. Vet.-Ak. Förh.* **19,** 210 (Rhyparochromida).

2. Geocorinæ (Stål) 1862, *Öfv. Vet.-Ak. Förh.* **19,** 212 (Geocorida).
3. Blissinæ (Stål) 1862, *Öfv. Vet.-Ak. Förh.* **19,** 10 (Blissida).
4. Cyminæ (Stål) 1862, *Öfv. Vet.-Ak. Förh.* **19,** 210 (Cymida).
5. Ischnorhynchinæ (Stål) 1872, *Öfv. Vet.-Ak. Forh.* **29,** 40 and 62 (Ischnorhyncharia).
6. Lygæinæ (Stål) 1862, *Öfv. Vet.-Ak. Förh.* **19,** 10 (Lygæida).

Syn. Astacopinæ Kirkaldy (*partim*) 1907, *Canadian Ent.* **39,** 244.

7. Orsillinæ (Stål) 1872, *Öfv. Vet.-Ak. Förh.* **29,** 43. (Orsillaria)
8. Oxycareninæ (Stål) 1862, *Öfv. Vet.-Ak. Förh.* **19,** 212 (Oxycarenida).
Syn. Anemopharina Berg 1879, *Hem. Argent.* 285.
9. Bledionotinæ Reuter 1878, *Ann. Soc. ent. Fr.* 144.
Syn. Pamphantinæ Barber and Bruner 1933, *Journ. N.Y. ent. Soc.* **41,** 532.
10. Malcinæ (Stål) 1866, *Hemipt. Afric.* **2,** 121 (Malcida).
11. Lipostemmatinæ (Berg) 1879, *Hem. Argent.* 288 (Lipostemmatina).
12. Henestarinæ (Douglas and Scott) 1865, *Brit. Hemipt. Heteroptera* 22 (Henestaridæ).
13. Pachygronthinæ (Stål) 1865, *Hemipt. Afric.* **2,** 121, 145 (Pachygronthida).
14. Heterogastrinæ (Stål) 1872, *Öfv. Vet.-Ak. Förh.* **29,** 40, 62 (Heterogastrina).
Syn. Phygadicidæ Douglas and Scott 1865, *Brit. Hemipt. Heteroptera* 21.
15. Chauliopinæ Breddin 1907, *Deutsch. ent. Zeit.* 40.
16. Artheneinæ (Stål) 1872, *Öfv. Vet.-Ak. Förh.* **29,** 38, 47 (Artheneina).
17. Phasmosomatinæ (Kiritshenko) 1938, *Trud. Zool. Inst. Baku* **8,** 117.
18. Henicocorinæ Woodward 1968, *Proc. Roy. ent. Soc. Lond.* (B) **37** (9–10); 125–132.

Family Thaumastellidæ Stys 1964, *Cas. Cs. Spol. ent.* (*Acta Soc. ent. Cechoslov.*) **61,** 3; 238–253.
Syn. Thaumastellinæ Seidenstucker 1960, *Stuttgarter Beitr. zur Naturkunde, Stuttgart* **38,** 1–4.

Family Idiostolidæ Stys 1964, *Acta Soc. ent. Cechoslov.* **61,** 238.
Syn. Idiostolinæ Scudder 1962, *Canad. Ent.* **94,** 1066–1069.

Family Pyrrhocoridæ (Amyot and Serville) 1843, *Hist. nat. Hémipt.* xxviii, 265 (Pyrrhocorides).
Syn. Pyrrhocoridæ Dohrn 1859, *Cat. Hemipt.* 36.
Syn. Astemmatidæ Spinola 1850, *Tav. Sin. Hem., ex Mem. Mat. Fis. Soc. Ital. Sci. Modena* **25,** 79 (1852).

Family Largidæ (Amyot and Serville) 1843, *Hist. nat. Hémipt.* xxxviii, 273 (Largides).
Syn. Largidæ Dohrn, *Cat. Hemipt.* 36.
Syn. Euryopthalminæ Van Duzee 1916, *Check-list Hemipt. America N. of Mexico* 24.

Family Piesmatidæ (Amyot and Serville) 1843, *Hist. nat. Hémipt.* xl (Piesmides).

Syn. Piesmidæ Spinola 1850, *Tav. Sin. Hem. ex Mem. mat. Fis. Soc. Ital. Sci. Modena* **25,** 84 (1852).

Syn. Zosmenidæ Dohrn 1859, *Cat. Hemipt.* 41.

Syn. Zosmeridæ Douglas and Scott 1865, *Brit. Hemipt. Heteroptera* 237.

Family Thaumastocoridæ Kirkaldy 1908, *Proc. Linn. Soc. N.S. Wales* **32** (corrigenda).

Syn. Thaumastotheriinæ Kirkaldy 1908, *Proc. Linn. Soc. N.S. Wales* **32,** 777.

Subfamily 1. Thaumastocorinæ Kirkaldy 1908, *Proc. Linn. Soc. N.S. Wales* **32,** 768–788.

2. Xylastodorinæ Barber 1920, *Bull. Brooklyn ent. Soc.* **15,** 98–105.

Syn. Discocorinæ Kormilev 1955, *Rev. Soc. Ent. Argent.* **18,** 7–10.

Family Berytidæ Fieber 1851, *Genera Hydroc.* 9.

Syn. Neididæ Kirkaldy 1902, *Journ. Bombay Nat. Hist. Soc.* **14,** 302.

Subfamily 1. Berytinæ Puton 1886, *Cat. Hemipt. Palæarct.* edn. 3, 19.

2. Metacanthinæ Douglas and Scott 1865, *Brit. Hemipt. Heteroptera* 99, 145.

Family Colobathristidæ (Stål) 1866, *Hemipt. Afric.* **2,** 121 (Colobathristida.

Family Aradidæ (Spinola) 1837, *Essai Hemipt.* 157 (Aradites).

Subfamily 1. Isoderminæ Stål 1872, *Svenska Vet.-Ak. Handl.* **10,** 4.

2. Prosympiestinæ Usinger and Matsuda 1959, *Class. Aradidæ, Brit. Mus.* 62.

3. Chinamyersiinæ Usinger and Matsuda 1959, *Class. Aradidæ, Brit. Mus.* 79.

4. Aradinæ (Amyot and Serville) 1843, *Hist. nat. Hémipt.* xli, 306 (Aradites).

5. Calisiinæ (Stål) 1873, *Svenska Vet.-Ak. Handl.* **11,** No. 2; 138 (Calisaria).

6. Aneurinæ (Douglas and Scott) 1865, *Brit. Hemipt. Heteroptera,* 26, 267 (Aneuridæ).

7. Carventinæ Usinger 1950, *VIIth Intern. Congr. Ent.* 176.

8. Mezirinæ Oshanin 1908, *Verz. Paläark. Hem.* **4,** 78.

Syn. Brachyrhynchinæ (Amyot and Serville) 1843, *Hist. nat. Hémipt.* xli, 303 (Brachyrhynchides).

Syn. Dysodiinæ Reuter 1912, *Öfv. Finska Vet. Soc. Förh.* **54a,** 33, 49, 57.

Syn. Chelonocorinæ Miller 1938, *Ann. Mag. nat. Hist.* (II), **1,** 498–510.

Family Termitaphididæ Myers 1924, *Psyche* **31,** 6; 267.

Syn. Termitocoridæ Silvestri 1911, *Portici Boll. Lab. Zool.* **5,** 231–236.

Family Joppeicidæ Reuter 1910, *Acta Soc. Sci. Fenn.* **37,** 3; 75.

Family Tingidæ (Costa A.) 1838, *Cimicum Regni Neap. Cent.* **1,** 18 (Tingini).

Syn. Tingidites Laporte 1832, *Essai Classif. Syst. Hem.* 47.

Syn. Tingidæ (Amyot and Serville) 1843, *Hist. nat. Hémipt.* xi, 295 (Tingides).

Syn. Tingitidæ aucct. (invalidated).

Subfamily 1. Tinginæ (Douglas and Scott) 1865, *Brit. Hemipt. Heteroptera,* 24 (Tingididæ).

2. Cantacaderinæ (Stål) 1873, *Enum. Hemipt.* **3,** 116 (Cantacaderaria).

3. Agramminæ (Douglas and Scott) 1865, *Brit. Hemipt. Heteroptera* 24, 242 (Agrammidæ).

Syn. Serenthiinæ Stål 1873, *Enum. Hemipt.* **3,** 116.

Family Vianaididæ Kormilev 1955, *Rev. Ecuat. Ent.* **2** (3–4); 465–477.

Family Enicocephalidæ (Stål) 1860, *Rio. Jan. Hemipt.* **1,** 81; *K. Svensk. Vet.-Ak. Handl.* **2,** No. 7, 1858.

Syn. Henicocephalidæ (Stål) 1865, *Hemipt. Afric.* **3,** 165 (Henicocephalida).

Subfamily 1. Enicocephalinæ Ashmead 1893, *Proc. Ent. Soc. Wash.* **2,** 328.

2. Aenictopechinæ Usinger 1932, *Pan Pacific Entomologist* **8,** 149.

Family Phymatidæ (Laporte) 1832, *Essai Classif. Syst. Hémipt.* 14 (Phymatites).

Syn. Phymatides (Amyot and Serville) 1843, *Hist. nat. Hémipt.,* xxxix, 288.

Syn. Macrocephalidæ Kirkaldy 1899, *Entomologist* **32,** 221.

Subfamily 1. Macrocephalinæ (Amyot and Serville) 1843, *Hist. nat. Hémipt.* xxxix, 291 (Macrocephalides).

Syn. Macrocephalinæ (Dohrn) 1859, *Cat. Hemipt.* 41 (Macrocephalidæ).

2. Phymatinæ (Dohrn) 1859, *Cat. Hemipt.* 41 (Phymatidæ).

3. Carcinocorinæ Handlirsch 1897, *Ann. K.K. Nat. Hofmus. Wien* **12,** 142.

4. Themonocorinæ (Carayon, Usinger and Wygodzinsky) 1958, *Rev. Zool. bot. Afr.* **57,** fasc. 3-4; 278 (Themonocorini).

Family Elasmodemidæ Letheirry and Severin 1896, *Cat. Hémipt.* **3,** 49.

Syn. Elasmocorinæ Usinger 1943, *Ann. ent. Soc. Amer.* **36,** 612.

Syn. Elasmodemidæ Wygodzinsky 1944, *Revue Brasil. Biol.* **4** (2); 205.

Family Reduviidæ Latreille 1807, *Gen. Crust. Ins.* **3,** 126.

Subfamily 1. Emesinæ (Amyot and Serville) 1843, *Hist. nat. Hémipt.* xlviii, 393 (Emesides).

Syn. Emesinæ Spinola 1850, *Tav. Sin. Hem.*, 45, *ex Mem. Mat. Fis. Soc. Ital. Sci. Modena* **25** (1852).

Syn. Ploiariinæ Costa 1852, *Cimic. Regni Neap. Cent.* **4,** 66.

Subfamily 2. Saicinæ (Stål) 1859, *Berlin ent. Zeit.* **3,** 328 (Saicida).

3. Visayanocorinæ Miller 1952, *Eos* **28,** 88–90.

4. Holoptilinæ (Amyot and Serville) 1843, *Hist. nat. Hémipt.* xlii, 318 (Holoptilides).

Syn. Holoptilinæ (Stål) 1859, *Berlin ent. Zeit.* **3,** 328 (Holoptilida).

5. Tribelocephalinæ (Stål) 1866, *Hemipt. Afric.* **3,** 44 (Tribelocephalida).

6. Bactrodinæ (Stål) 1866, *Hemipt. Afric.* **3,** 45 (Bactrodida).

7. Stenopodinæ (Amyot and Serville), *Hist. nat. Hémipt.* xlviii, 386 (Stenopodides).

Syn. Stenopodinæ (Stål), 1859, *Berlin ent. zeit.* **3,** 328 (Stenopodida).

8. Salyavatinæ (Amyot and Serville) 1843, *Hist. nat. Hémipt.* xliv, 349 (Salyavatides).

Syn. Salyavatinæ (Stål) 1859, *Berlin ent. zeit.* **3,** 328 (Salyavatida).

9. Sphaeridopinæ (Amyot and Serville) 1843, *Hist. nat. Hémipt.*, xlvii, 381 (Sphaeridopides).

Syn. Sphaeridopinæ (Pinto) 1927, *Bol. Biol. S. Paulo* **6,** 43, 47 (Sphæridopidæ).

10. Manangocorinæ Miller 1954, *Idea*, **10,** 2.

11. Physoderinæ Miller 1954, *Tijdsch. v. Ent.* **96,** 82.

12. Centrocneminæ Miller 1956, *Bull. Brit. Mus. Ent.* **4,** 219–283.

13. Chryxinæ Champion 1898, *Biol. Centr. Amer. Rhynchota, Het.* **2,** 180.

14. Vesciinæ Fracker and Bruner 1924, *Ann. Soc. ent. Amer.* **17,** 165.

Syn. Chopardititæ Villiers 1944, *Bull. Soc. ent. Fr.* **49,** 79.

15. Cetherinæ Jeannel 1919, *Voy. Alluaud Jeannel Afr. or. Hém.*, 1911–1912, 178.

16. Eupheninæ Miller 1955, *Ann. Mag. nat. Hist.* (12), **8,** 449–452.
17. Reduviinæ (Amyot and Serville) 1843, *Hist. nat. Hémipt.* xliii, 333 (Reduviides).
Syn. Reduviinæ Spinola 1850, *Tav. Sin. Hem.* 145, *ex Mem. mat. Fis. Soc. Ital. Sci., Modena* **25** (1852).
Syn. Acanthaspinæ (Stål) 1866, *Hemipt. Afric.* **3,** 44 (Acanthaspidida).
18. Triatominæ Jeannel 1919, *Voy. Alluaud Jeannel Afr. or. Hém.* 176, 177.
Syn. Conorhininæ (Amyot and Serville) 1843, *Hist. nat. Hémipt.* xlviii, 383 (Conorhinides).
19. Piratinæ (Stål) 1859, *Berlin ent. Zeit.* **3,** 328 (Peiratida).
20. Phimophorinæ Handlirsch 1897, *Verh. zool. bot. Ges. Wien* **47,** 408.
21. Mendanocorinæ Miller 1956, *Ann. Mag. nat. Hist.* (12) **9,** 587–589.
22. Hammacerinæ (Stål) 1859, *Berlin ent. Zeit.* **3,** 328 (Hammacerida).
Syn. Hammatocerinæ (Stål) 1862, *Stett. ent. Zeit.* **23,** 455 (Hammatocerida).
Syn. Microtominæ Schumacher 1924, *Deutsch. ent. Zeit.* 336.
23. Ectrichodiinæ (Amyot and Serville) 1843, *Hist. nat. Hémipt.*, xliv, 342 (Ectrichodides).
Syn. Ectrichodiinæ Spinola 1850, *Tav. Sin. Hem.* 44, 45, *ex Mem. Mat. Fis. Soc. Ital. Sci. Modena* **25** (1852).
24. Rhaphidosomatinæ Jeannel 1919, *Voy. Alluaud Jeannel Afr. or. Hém.* 263.
25. Harpactorinæ (Amyot and Serville) 1843, *Hist. nat. Hémipt.* xlv, 355 (Harpactorides).
Syn. Harpactorinæ Spinola 1850, *Tav. Sin. Hem.* 45, *ex Mem. Mat. Fis. Soc. Ital. Sci., Modena* **25** (1852).
Syn. Reduviinæ (Stål) 1859, *Öfv. Vet.-Ak. Förh.* **16,** 19 (Reduvides).
26. Apiomerinæ (Amyot and Serville) 1843, *Hist. nat. Hémipt.* xliv, 350 (Apiomerides).
Syn. Apiomerinæ (Stål) 1859, *Berlin ent. zeit.* **3,** 328 (Apiomerida).
27. Ectinoderinæ (Stål) 1866, *Öfv. Vet. Ak. Förh.* **23,** 245 (Ectinoderida).
28. Phonolibinæ Miller, *Eos*, **28,** 86.
29. Perissorhynchinæ Miller 1952, *Eos* **28,** 87.
30. Tegeinæ Villiers 1948, *Hémipt. Réduv. Afr. noire*, 171.

31. Diaspidiinæ Miller 1959, *Bull. Brit. Mus., Ent.* **8,** 2; 103.

Family Pachynomidæ (Stål) 1873, *Enum. Hemipt.* **3,** 107 (Pachynomina).
Syn. Pachynomidæ Carayon 1954, *Bull. Soc. Zool. France* **79,** 191 (as a family).

Family Velocipedidæ Bergroth 1891, *Wien ent. Zeit.* **10,** 265.
Syn. Scotomedinæ Blöte 1945, *zool. Meded.* **25,** 323 (as subfa mily of Nabidæ).

Family Medocostidæ Stys 1967, *Acta ent. bohemoslov* **64,** 439–465.

Family Nabidæ Costa 1852, *Cimic. Regni Neap. Cent.* **3,** 66.

Subfamily 1. Nabinæ Reuter 1890, *Rev. Ent.* **9,** 293.
Syn. Reduviolinæ Reuter and Poppius 1909 (not Reuter 1890), *Acta Soc. Sci. Fenn.* **37,** 2; **3.**
Syn. Coriscinæ (Stål) 1873, *Enum. Hemipt.,* **3,** 106 (Coriscina).
2. Prostemminæ Reuter 1890, *Rev. Ent.* **9,** 289.
Syn. Nabinæ Reuter and Poppius 1909 (not Reuter 1890), *Acta Soc. Sci. Fenn.* **37,** 2, 3.
3. Arachnocorinæ Reuter 1890, *Rev. Ent.* **9,** 292.
4. Gorpinæ Reuter 1909, *Ann. Soc. ent. belg.* **53,** 423.
5. Carthasinæ Blatchley 1926, *Het. East. N. Amer.* 538–539.

Family Polyctenidæ Westwood 1874, *Thesaur. Ent.* 197.

Family Cimicidæ (Latreille) 1804, *Hist. nat. Crust. Ins.* **12,** 235 (Cimicides).
Syn. Cimicidæ (Leach) 1815, *Brewster's Edinb. Encyc.* **9,** 122 (Cimicida).
Syn. Acanthiadæ Fieber 1861, *Europ. Hemipt.* 37, 135.
Syn. Acanthiidæ Douglas and Scott 1865 (nec Leach 1815), *Brit. Hemipt. Heteroptera*, 37.
Syn. Clinocoridæ Kirkaldy 1906, *Trans. Amer. ent. Soc.* **32,** 147.

Subfamily 1. Cimicinæ Van Duzee 1916, *Check-list Hemipt. Amer. N. of Mexico* 33.
2. Haematosiphoninæ Jordan and Rothschild 1912, *Novitat. Zool.* **19,** 352.
3. Cacodminæ Kirkaldy 1899, *Bull. Liverpool Mus.* **2,** 45.
4. Primicimicinæ Usinger, R. L. *et al.* 1966, *Monograph of Cimicidæ* (*Hemiptera-Heteroptera*) *Thomas Say Foundation*, Vol. VII.

Family Anthocoridæ (Amyot and Serville) 1843, *Hist. nat. Hémipt.* xxxvii, 262 (Anthocorides).
Syn. Anthocoridæ Fieber 1851, *Genera Hydroc.* 9.

Subfamily 1. Lyctocorinæ Reuter 1884, *Monog. Anthoc. Acta Soc. Sci. Fenn.* **14** (1885), 558.
2. Anthocorinæ Reuter 1884, *Monog. Anthoc., Acta Soc. Sci. Fenn.* **14** (1885), 558.
3. Dufouriellinæ Van Duzee 1916, *Check-list Hemipt. Amer. N. of Mexico*, 35.
Syn. Xylocorinæ Reuter 1884, *Monog. Anthoc. Acta Soc. Sci. Fenn.* **14** (1885), 558.

Family Microphysidæ Dohrn 1859, *Cat. Hemipt.* 36.

Family Plokiophilidæ (China) 1953, *Ann. Mag. nat. Hist.* (12), **6,** 73.

Subfamily 1. Plokiophilinæ China 1953, *Ann. Mag. nat. Hist.* (12) **6,** 73.
2. Embiophilinæ Carayon 1960, *Verh. XI, Int. Kongress Ent. Wien* **1,** 711–714.

Family Miridæ (Hahn) 1831, *Wanz. Ins.* **1,** 234 (Mirides).
Syn. Capsidæ (Burmeister) 1835, *Handb. Ent.* **2,** 263 (Capsini).
Syn. Phytocoridæ Fieber 1858, *Wien ent. Monatschrift* **2,** 289.

Subfamily 1. Mirinæ (Hahn) 1833, *Wanz. Ins.* **1,** 234 (Mirides).
Syn. Mirinæ (Amyot and Serville) 1843, *Hist. nat. Hémipt.* xxxviii, 277 (Mirides).
Syn. Mirinæ (Reuter) 1910, *Acta Soc. Sci. Fenn.* **37,** 109, 128, 155 (Mirina).
2. Orthotylinæ Van Duzee 1916, *Check-list Hemipt. Amer. N. of Mexico*, 203.
Syn. Heterotominæ (Reuter) 1910, *Acta Soc. Sci. Fenn.* **37,** 3; 114 (Heterotomina).
Syn. Cyllecorinæ Oshanin 1912, *Kat. Paläarkt. Hemipt.* 72.
3. Phylinæ (Douglas and Scott) 1865, *Brit. Hemipt. Heteroptera* 30, 346 (Phylidæ).
Syn. Plagiognathinæ Oshanin 1912, *Kat. Paläarkt. Hemipt.* 77.
4. Bryocorinæ (Baerensprung) 1860, *Cat. Hem. Eur.* 13 (Bryocorides).
Syn. Bryocorinæ (Douglas and Scott) 1865, *Brit. Hemipt. Heteroptera* 28, 276 (Bryocoridæ).
5. Deræocorinæ (Douglas and Scott) 1865, *Brit. Hemipt. Heteroptera* 29, 315 (Deræocoridæ).
Syn. Termatophylidæ (Reuter) 1888, *Wien ent. Zeit.* **3,** 218 (Termatophylina).
Syn. Cliveneminæ (Reuter) 1875, *Caps. Bor. Amer.* 62; *Öfv. Vet.-Ak. Forh.* **32,** No. 9; 54–92 (Clivenemaria).
Syn. Hyaliodinæ Knight 1943, *Ent. News* **54** (5), 19.
Syn. Ambraciinæ (Reuter) 1910, *Acta Soc. Sci. Fenn.* **37,** 109, 154 (Ambraciina).

6. Cylapinæ Kirkaldy 1903, *Wien ent. Zeit.* **22,** 13.
Syn. Bothynotinæ (Reuter) 1910, *Acta. Soc. Sci. Fenn.* **37,** 109, 155 (Bothynotina).
Syn. Lygæoscytinæ (Reuter) 1910, *Acta Soc. Sci. Fenn.* **37,** 110 (Lygæoscytina).

Family Isometopidæ Fieber 1860, *Wien ent. Monat.* **4,** 259.

Family Dipsocoridæ Dohrn 1859, *Cat. Hemipt.* 36.
Syn. Cryptostemmatidæ McAtee and Malloch 1925, *Proc. U.S. Nat. Mus.* **17,** 1.
Syn. Ceratocombidæ Fieber 1860, *Europ. Hemipt.* 25, 39, 142.

Family Schizopteridæ (Reuter) 1891, *Acta Soc. Sci. Fenn.* **19,** 6; 3 (Schizopterina).

Family Hydrometridæ (Billberg) 1820, *Enum. Ins. Mus. Billb.* 67 (Hydrometrides).
Syn. Limnobatidæ Fieber 1860, *Europ. Hemipt.* 23.
Subfamily 1. Hydrometrinæ Esaki 1927, *Entomologist* **60,** 4.
2. Limnobatodinæ Esaki 1927, *Entomologist* **60,** 4.
3. Heterocleptinæ Villiers 1948, *Réduv. Afr. noire* 174 (described in the Reduviidæ).
Syn. Hydrobatodinæ China and Usinger 1949, *Rev. Zool. Bot. Afr.* **41,** 4; 318.

Family Gerridæ Leach 1815, *Brewster's Edinb. Encyc.* **9,** 123.
Syn. Gerridæ (Amyot and Serville) 1843, *Hist. nat. Hémipt.* 1, 410 (Gerrides).
Syn. Hydrometridæ (Fieber) 1860, *Europ. Hemipt.* 24 (Hydrometræ).
Subfamily 1. Gerrinæ Bianchi 1896, *Ann. Mus. Petersb.* 69.
2. Halobatinæ Bianchi 1896, *Ann. Mus. Petersb.* 1896, 69.
3. Hermatobatinæ Coutière and Martin 1901, *S.R. Acad. Sci. Paris,* **132,** 1066–1068.
4. Rhagadotarsinæ Lundblad 1933, *Archiv. für Hydrobiol. Suppl.* **12,** 411.
5. Ptilomerinæ Esaki 1927, *Eos* **2,** 252.

Family Veliidæ (Amyot and Serville) 1843, *Hist. nat. Hémipt.* 1, 418 (Velides).
Syn. Veliidæ Dohrn 1859, *Cat. Hemipt.* 53.
Syn. Veliidæ (Reuter) 1912, *Öfv. Finska Vet.-Soc. Förh.* **6,** 14, 18.
Subfamily 1. Perittopinæ China and Usinger 1949, *Ann. Mag. nat. Hist.* (12) **2,** 350.
2. Rhagoveliinæ China and Usinger 1949, *Ann. Mag. nat. Hist.* (12) **2,** 350.
3. Hebroveliinæ (Lundblad) 1939, *Ent. Tidsk.* **60,** 1–2, 33 (Hebroveliidæ).

4. Hydroessinæ (Fieber) 1860, *Europ. Hemipt.* 23 (Hydroessæ).
Syn. Microveliinæ China and Usinger 1949, *Ann. Mag. nat. Hist.* (12) **2,** 351.
5. Veliinæ China and Usinger 1949, *Ann. Mag. nat. Hist.* (12) **2,** 351.
6. Haloveliinæ Esaki 1930, *Journ. F.M.S. Mus.* **16,** 22.
Syn. Haloveliidæ Poisson 1956, *Mem. Inst. Sci. Madagascar* (E), **7,** 255.

Family Mesoveliidæ Douglas and Scott 1867, *Ent. mon. Mag.* **4,** 3.
Syn. Mesoveliadæ Reuter 1912, *Öfv. Finska Vet. Soc. Förh.* **6,** 17, 23, 47, 49.
Subfamily 1. Mesoveliinæ Usinger.
2. Mesoveloideinæ Usinger.
3. Macroveliinæ (McKinstry) 1942, *Pan. Pacif. Ent.* **18,** 91 (Macroveliidæ).

Family Hebridæ (Amyot and Serville) 1843, *Hist. nat. Hemipt.* xl, 293 (Hebrides).
Syn. Næogeidæ Kirkaldy 1902, *Faun. Hawaii,* **3,** 168.

Family Leotichiidæ China 1933, *Ann. Mag. nat. Hist.* (10) **12,** 185.

Family Leptopodidæ Costa 1838, *Cimic. Regni Neap. Cent. I, Atti real. Ist. incorragg. alle Sci. nat. Nap.* **7,** 151 (1847).

Family Saldidæ (Amyot and Serville) 1843, *Hist. nat. Hémipt.* xlix (Saldides).
Syn. Saldidæ Costa 1852, *Cimic. Regni. Neap. Centr.* 3, *Atti real. Ist. incorragg. alle Sci. nat. Nap.* **8,** 66 (1855).
Syn. Acanthiidæ (Leach) 1815, *Brewster's Edinb. Encyc.* **9,** 123 (Acanthides).
Syn. Acanthiadæ Fieber 1860, *Europ. Hemipt.* 24.
Subfamily 1. Saldinæ Van Duzee 1917, *Cat. Hemipt. America North of Mexico* 438.
2. Saldoidinæ Reuter 1912, *Öfv. Finska-Vet. Soc. Förh.* **54A,** No. 12; 23.
3. Aepophilinæ (Puton) 1879, *Synop. Hém. Hét. France* **2,** 145 (Aepophilidæ).

Family Notonectidæ (Leach) 1815, *Brewster's Edinb. Encyc.* **9,** 124 (Notonectida).
Subfamily 1. Anisopinæ Hutchinson 1929, *Ann. S. Afr. Mus.* **25,** 3; 362.
2. Notonectinæ (Leach) 1815, *Brewster's Edinb. Encyc.* **9,** 124 (Notonectida).

Family Pleidæ (Fieber) 1851, *Genera Hydroc.* 27 (Pleæ).

Family Helotrephidæ Esaki and China 1927, *Trans. ent. Soc. Lond.* 280.

Subfamily 1. Neotrephinæ China 1940, *Ann. Mag. nat. Hist.* (II) **5,** 123.
2. Idiocorinæ Esaki and China 1927, *Trans. ent. Soc. Lond.* 280.
3. Helotrephinæ Esaki and China 1927, *Trans. ent. Soc. Lond.* 280.

Family Corixidæ (Leach) 1815, *Brewster's Edinb. Encyc.* **9,** 124 (Corixida).
Syn. Corixidæ (Amyot and Serville 1843, *Hist. nat. Hémipt.* li, 444 (Corisides).

Subfamily 1. Micronectinæ Jaczewski 1924, *Ann. Zool. Mus. Polon. Hist. nat.* **3,** 3.
Syn. Sigaridæ Douglas and Scott 1865, *Brit. Hemipt. Heteroptera* 50.
2. Diaprepocorinæ Lundblad 1928, *Entom. Tidsk.* **1,** 9.
3. Corixinæ (Douglas and Scott) 1865, *Brit. Hemipt. Heteroptera*, 50 (Corixidæ).
4. Stenocorixinæ Hungerford 1948, *Univ. Kansas Sci. Bull.* **32,** 43.
5. Cymatiinæ Walton 1940, *Trans. Connect. Acad. Arts Sci.* **33,** 344.
6. Heterocorixinæ Hungerford 1948, *Univ. Kansas Sci. Bull.* **32,** 43.

Family Nepidæ (Latreille) 1802, *Hist. nat. Crust. Ins.* **3,** 252 (Neparïæ).

Subfamily 1. Nepinæ (Douglas and Scott) 1865, *Brit. Hemipt. Heteroptera*, 47 (Nepidæ).
2. Ranatrinæ (Douglas and Scott) 1865, *Brit. Hemipt. Heteroptera*, 46 (Ranatridæ).

Family Belostomatidæ (Leach) 1815, *Brewster's Edinb. Encyc.* **9,** 123 (Belostomida).
Syn. Belostomidæ Dohrn 1859, *Cat. Hemipt.* **54**.

Family Naucoridæ Fallèn 1814, *Spec. Nov. Disp. Meth.* 3, 15.

Subfamily 1. Naucorinæ Stål 1876, *Enum. Hemipt.* **5,** 142.
2. Limnocorinæ Stål 1876, *Enum. Hemipt.* **5,** 142.
3. Laccocorinæ Stål 1876, *Enum. Hemipt.* **5,** 142.
4. Cryphocricinæ Montandon 1879, *Verh. Zool. Bot. Ges. Wien* **47,** 6.
5. Ambrysinæ Usinger 1941, *Ann. ent. Soc. Amer.* **34,** 911.
6. Cheirochelinæ Montandon 1897, *Ann. Mus. Civ. Genova* **37,** 367.
7. Potamocorinæ Usingeı 1941, *Ann. ent. Soc. Amer.* **34,** 89.

8. Aphelocheirinæ (Fieber) 1860, *Europ. Hemipt.* 23 (Aphelochiræ).

Syn. Aphelocheiridæ aucct.

Family Gelastocoridæ Kirkaldy 1897, *Entomologist* **30,** 258.

Syn. Galgulidæ (Billberg) 1820, *Enum. Ins. Mus. Billb.* 66 (Galgulides).

Subfamily 1. Gelastocorinæ Champion 1901, *Biol. Centr. Amer. Rhynchota, Het.* **2,** 437.

2. Nerthrinæ (Kirkaldy) 1906, *Trans. ent. Soc. Amer.* **32,** 149 (Nerthridæ).

Syn. Mononychinæ (Fieber) 1851, *Genera Hydroc.* 9, 12 (Mononycoidea).

Family Ochteridæ Kirkaldy 1906, *Trans. Amer. ent. Soc.* **32,** 149.

Syn. Pelogoniidæ (Leach) 1815, *Brewster's Edinb. Encyc.* **9,** 123 (Pelogonida).

Chapter 2

Development

A great diversity of form is exhibited by the ova of the Heteroptera and, so far as our knowledge goes, each group of genera in a subfamily would appear to have a characteristic ovum. It is not correct to say that ova are of a definite type in each subfamily.

The principal constituent of an ovum is the shell or chorion which is formed by a cuticular secretion of the epithelial cells in the oviduct. Generally it is made up of two layers of different thickness and structure and the external surface may be smooth and shining, shagreened or sculptured in one manner or another.

Some Heteropterous ova, notably those of certain Pentatomidæ, have a series of short, hollow appendages, usually somewhat wider at the apex, situated on the upper margin of the chorion. Ova bearing appendages of this type are produced by certain Coreidæ, Urostylidæ, Reduviidæ and Miridæ.

When these appendages were first noticed and subsequently more closely examined, the conclusion arrived at was that it was through them that the sperm entered the ovum. This was the opinion of Leuckart who advanced it in a treatise on the eggs of insects, but, about sixty years later, Gross disputed it and suggested that these chorionic processes (Chorion-Anhänge) were for the purpose of ventilating the interior of the ovum.

At the present time it seems to be the opinion generally accepted but, it must be noted that although all ova require air, there are very many, in fact the majority which are not provided with these processes. Aeration in such cases is effected through pores in the operculum or through the chorion itself.

Certain species secrete a glutinous substance which envelops each ovum at the time of oviposition, therefore air can penetrate only through pores in the operculum which is commonly a circular concavo-convex plate but occasionally has a very complicated structure (cf. *Rhinocoris* spp. Reduviidæ-Harpactorinæ).

Viviparity has been reported in certain species and is more widely spread than one has thought, up to the present. It is fairly varied and comprises cases in which neanides almost completely formed and even partly free from the chorion have been found in some females. This type may be termed oviparity and has been recorded in the case of some females of *Loricula pselaphiformis* (Curt.) 1833 (Microphysidæ) and *Aneurus madagascariensis* Hoberl. 1957 (Aradidæ), and also in the **Plokiophilinæ** China 1953.

Many methods of oviposition are met with in the Heteroptera; some species deposit their ova in groups, others singly, either with or without a small quantity of glutinous substance to fix them to the substratum.

Then there are those species which insert their ova into the soil or into the softer parts of plants. Aquatic species immerse their ova and attach them to plants or stones as a rule. Exceptionally, the dorsum of the male of the same species is selected by the female (cf. certain Belostomatidæ).

The position of the ovum in relation to the substratum varies, some ova being deposited with the longer axis vertical, others at an angle or horizontal. Some ova which have the horizontal position have a short pedicel which raises them above the surface of the object on which they are deposited.

The construction of an ootheca such as is found in the Blattidæ or Mantidæ (Orthoptera) is not met with. The nearest approach to that type of oviposition is that of certain **Harpactorinæ** (Reduviidæ), namely species of *Sycanus* Amyot and Serville 1843, in which the ova, after deposition in a vertical position in groups, are covered by the female with a glutinous substance which envelops all of them with the exception of much of the differentiated portion of the chorion. Another Harpactorine *Isyndus heros* (Fabricius) 1803, secretes a substance which takes the form of bubbles which harden and form together a structure which entirely covers the ova with the exception of the opercula (Fig. 45, 10). In *Panthous* Stål 1863, of the same subfamily, the ova are completely covered with a somewhat tenuous glutinous substance. In the case of both *Sycanus* and *Panthous* it must be pointed out that the glutinous substance does not become hard.

In the chapters in which the families are dealt with individually, the various methods of oviposition are described more fully.

At the time of eclosion the chorion may be split in various directions by the embryo, but if there is an operculum, this alone is removed.

To assist the embryo to remove the operculum an egg-burster (Eisprenger), a highly sclerotized portion of the embryonic cuticle, present in the Pentatomidæ, Coreidæ and in other families is employed.

The egg-burster has various shapes (Fig. 1). In the Pentatomidæ it is commonly anchor-shaped or has the form of a T with a short, conical spur at one or both ends of the 'shaft' portion, or it may be an elongate rod with the middle portion thickened.

In the case of the Pentatomidæ, when eclosion is taking place, one or both spurs engage with the rim of the operculum, so that the latter when raised, may not close again when the embryo retracts prior to distending itself once more to lift the operculum higher.

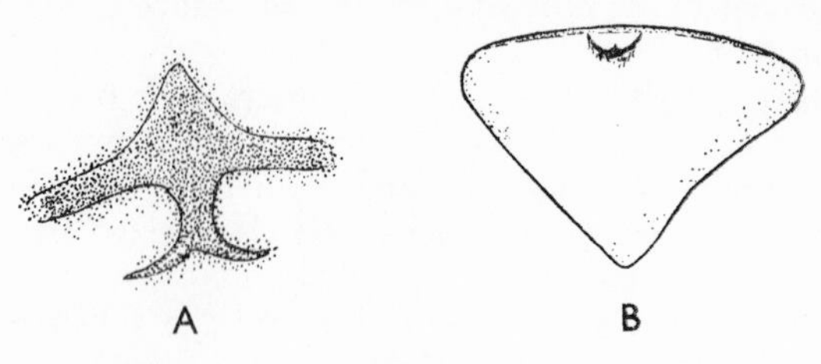

Fig. 1
Egg-bursters. A. *Embolosterna taurus* Westwood 1837 (Pentatomidæ-Tessaratominæ). B. *Physomerus parvulus* Dallas 1851 (Coreidæ-Coreinæ).

The initial split in the chorion is made by the egg-burster of Coreidæ, but after that it does not render further assistance to the embryo disengaging itself from the chorion.

An analogous organ is present in the Reduviidæ, but the purpose is not exactly the same as in the instances just mentioned, that is to say to pierce the chorion. It consists of two groups of very small denticles, more strongly sclerotized parts of the embryonic cuticle which, it would seem, engage with the lower surface of the operculum to facilitate its removal. When the operculum is sufficiently raised, or when the gap between it and the chorion has been sufficiently widened, the embryo, by even greater distension of the body caused the enveloping cuticle to split and is then able to free itself completely. The operculum and the cuticle often remain on the empty chorion.

An egg-burster has been discovered in the Cydnids, *Sehirus bicolor* (Linnæus) 1758, and *Corimelæna virilis* McAtee and Malloch 1933. It is present also in the ovum of *Cænocoris* sp. (Lygæidæ-Lygæinæ) and consists of a relatively large more highly sclerotized portion of the embryonic cuticle with a still more strongly sclerotized transverse ridge.

In the ova of Cimicidæ also an egg-burster has been noted. This consists of a V-shaped series of spines on the cuticle, apparently similar in shape but not similarly arranged as in the Reduviid ova.

So far as observations go, newly hatched neanides remain on or near the empty ova until they moult for the first time. During that period they feed very little or not at all.

Neanides from ova deposited in groups are often gregarious during the first instar, but disperse after the first moult. In some species of Coreidæ, Aradidæ and Pyrrhocoridæ which often oviposit on isolated plants or in a restricted area, the neanides remain together until they reach the adult stage.

Examples of this gregarious habit may be seen in the Coreids, *Petascelis remipes* Signoret 1847, an Ethiopian species which feeds on wild *Gardenia*, *Physomerus grossipes* (Fabricius) 1794 (**Coreinæ**), an abundant species in Malaya, which feeds on Convolvulaceæ and Leguminosæ, the Pyrrhocorids *Dysdercus* Amyot and Serville 1843, represented in the Ethiopian and Oriental Regions on many plants including cotton and other Malvaceæ, and it is interesting to note that the adults remain with the neanides also. It is possible, therefore, to see all stages of these bugs together at one time, with the exception, of course, of *Dysdercus* which oviposit in the soil.

The Pyrrhocorid *Melamphaus faber* (Fabricius) 1787, has been observed to appear in a very large swarm on one occasion in Malaya, but the individuals forming the swarm dispersed within a short time, apparently after oviposition had been completed. At most times of the year this species is widely distributed in jungle areas and consequently appears to be uncommon.

At the time of ecdysis the neanide attaches itself as firmly as possible to the substratum with its tarsal claws and frequently takes up a position in which it hangs head downwards. By distending itself with air it causes the cuticle to split along the so-called ecdysial line, a narrow, feebly sclerotized part located between the eyes, extending along the postocular portion of the head and along the nota to the first abdominal segment.

By successive convulsive movements the neanide then extricates the antennæ, head and anterior legs, and with these legs obtains a hold on some object close by. It then drags out the remaining legs and the body. Sometimes it relies on gravity to assist it to quit the exuviæ.

After ecdysis, a neanide remains quiescent for some time until the integument is hardened sufficiently. The period elapsing before the integument is hard and firm may cover several days. The development of the pigment as a rule takes less time, but it may be less rapid in those species which live where very little light penetrates.

From the time of eclosion to maturity, neanides usually pass through five or six instars and at each moult some modification, slight in the early instars but greater in succeeding ones, takes place in the external appearance. The modification concerns mainly the hemelytra and the metathoracic wings, the rudiments of which are plainly visible generally in the fourth and fifth instars, but less so in the third instar. Other modifications comprise an increase, in certain cases, in the number of tarsal segments, an increase in the number of antennal segments and the appearance of ocelli.

When the number of antennal sements is more than four in the adult, as, for example, in most genera of the Pentatomidæ, the second segment of the antennæ apparently divides, thus increasing the number to five. In the subfamily **Ectrichodiinæ** (Reduviidæ), the antennæ of the adults may have four, six, seven or eight segments the neanides having four segments. The increase in this instance can be seen to be taking place at the fourth and fifth instars, in the apical segment.

An increase in the number of antennal segments in species of *Opistoplatys* Westwood 1834 (Revudiidæ-Tribelocephalinæ), is probably caused by the division of the apical segment, but there is no information regarding this since the neanidal stages of this genus and, indeed, of any genera in this subfamily, have been studied hardly at all.

In the **Hammacerinæ** (Reduviidæ), represented in South America, the number of antennal segments in the adults may be more than forty, *e.g. Hammacerus cinctipes* Stål 1858, or over thirty, *e.g. H. luctuosus* Stål 1854, but, curiously enough, the division into so many segments is restricted not to the apical segments but to the second segment.

Division into this large number of segments which will eventually take place is more or less clearly indicated in the neanide of the fifth instar in *H. gayi* Spinola 1852, a specimen of which I have examined. In species belonging to the genus *Homalocerus* Perty 1833 of the same subfamily, namely *maculicollis* Stål 1872, *varius* Perty 1833 and *binotatus* Champion 1899, there are approximately fifteen, seven and ten segments respectively forming the second segment.

Another instance of antennal modification as regards the form and not the number may be seen in the neanides of *Cletus* Stål 1859 (Coreidæ-Coreinæ), which have the second and third segments expanded and flattened, but in the adult the segments are normal.

Other modifications which take place comprise the reduction in size or disappearance of spines or secretory hairs, the latter a characteristic of certain Reduviidæ.

Neanides sometimes possess remarkable structures which are lost in the final moult. An instance of this is the Mirid *Paracarnus myersi* China 1931 (Miridæ-Deræocorinæ) (Fig. 2) and *Hoffmannocoris chinai* Miller 1950 (Reduviidæ-Rhaphidosomatinæ) (Fig. 47).

The legs also undergo a considerable alteration in form, especially the anterior pair, the tibiæ of which in the neanides may be greatly expanded and become progressively smaller in each instar and are normal in the adult, cf. *Anoplocnemis phasiana* (Fabricius) 1781, *Ochrochira rubrotincta* Miller 1931 (Coreidæ-Coreinæ). Conversely, the neanide may have more or less normal anterior tibiæ which become expanded to a lesser degree in the adult, cf. *Petalochirus* Palisot Beauvois 1805 (Reduviidæ-Salyavatinae).

Legs which are strongly spinose in the neanide become less strongly so in the adult in certain cases, cf. *Hoffmannocoris chinai*. In *Scipinia spinigera* Reuter 1881, a Javanese species (Reduviidæ-Harpactorinæ) the spines on the anterior tibiæ develop at the fifth instar.

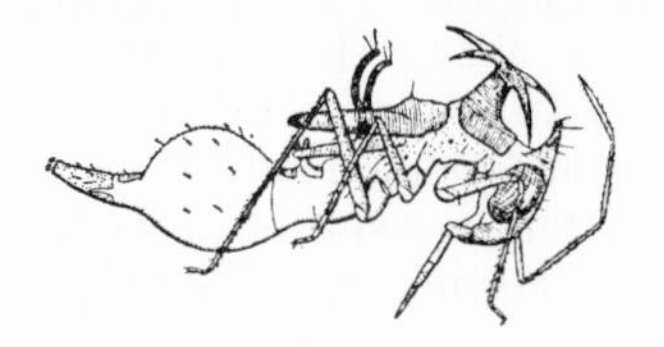

Fig. 2

Neanide of *Paracarnus myersi* China 1931. Miridæ-Deræocorinæ (original, China).

The *fossula spongiosa* present in the adults, and in some cases in the neanides of many Reduviidæ appear usually at the fourth instar. The tarsi, usually composed of two segments in the neanides, increase to three in most cases in the adult stage.

With regard to the development of the hemelytra and metathoracic wings, it is usual for the rudiments to appear at the third instar, and at the fifth instar they are often long enough to cover a considerable portion of the abdomen. In genera in which both apterous and alate adult forms occur, it is not always possible to decide when the neanide has reached the fifth instar whether the resulting adult will have wings or not, cf. *Maraenaspis* spp. (Reduviidæ-Ectrichodiinæ).

Two or more forms exhibiting different degrees of wing development are to be found in several families. Thus the adult may have fully developed wings (macropterous), have shortened wings (brachypterous or micropterous) or may be wingless (apterous).

In certain Reduviid genera, namely *Edocla* Stål 1859, *Paredocla* Jeannel 1914 (**Reduviinæ**), and *Maraenaspis* Karsch 1892 (**Ectrichodiinæ**), the males may be alate or apterous. Generally the metathoracic wings are shorter than the hemelytra, and in the case of brachyptery, the former are reduced proportionally.

Alary polymorphism is associated in the Heteroptera with modifications of the thoracic segments, chiefly the pronotum, reduction in size or complete absence of ocelli and occasionally a tendency to physogastry.

It is not uncommon for apterous individuals to be mistaken for immature forms. To be able to decide, it is essential first to examine the apical abdominal segments. Other adult characters will be found in the number and structure of the antennal segments and in the absence of dorsal abdominal gland ostioles. In the adult these are indicated only by a feeble modification of the segmental margins mid-dorsally of certain segments.

Characteristic of the neanides of many families are the abdominal glands, the ostioles of which are situated at the basal margins of the fourth to sixth dorsal segments, sometimes on the fourth and fifth only or solely on the sixth. These glands, to which the names 'repugnatorial', 'stink' or 'odoriferous' are given on account of the odorous and volatile fluid they secrete—which apparently assures the possessor of a certain amount of protection against enemies—are situated below the integument of the abdomen dorsally.

It has been observed that neanides are able to project the fluid for some distance, but generally it flows from the gland and spreads over the insect until it has volatilized, e.g. *Ochrochira rubrotincta.*

In the adults the glands are situated in the thorax with the ostiole between the meso- and metapleuron or in the metasternal depression adjacent to the inner margin of the acetabulum. When the ostiole is located between the meso- and metapleuron there is usually an evaporative area surrounding it, and the margins may be enlarged. The evaporative area is differently sculptured from the remainder of the segments, being usually sub-shagreened and with numerous shallow sulci. In the Pentatomid genus *Tyomana* Miller 1952, the ostiole of the metathoracic glands is hardly distinguishable from the surrounding puncturation, a difficulty that is enhanced by the fact that there is no evaporative area.

In the Reduviidæ the 'stink' glands in *Rhodnius prolixus* Stål 1859, and in *Triatoma rubrofasciata* (de Geer) 1773 **(Triatominæ),** were first discovered by Brindley. There are, however, two types of glands, those which are found in other families of Heteroptera, as well as in the Reduviidæ and another type situated laterally at the base of the abdomen and under the first abdominal tergite and which have two small ostioles situated near the posterior angle of the metathoracic epimeron. This type of ostiole has been noted in the genera *Eupheno* Gistel **(Eupheninæ),** *Cethera* Amyot and Serville 1843, *Cetheromma* Jeannel 1917, *Caprocethera* Breddin 1903 **(Cetherinæ)** and *Centrocnemis* Signoret 1852 **(Centrocneminæ).** In *Eupheno*, however, the ostiole is situated on a somewhat complicated and lamellar elevation which forms the evaporative area.

The odour of the volatile fluid to the human sense of smell is generally repellent, but it is not always so, on account of its similarity to that of pineapple, cinnamon, ether, for example. The odour of the secretion from some Reduviidæ resembles that of valerianic acid.

The value of the ability to secrete a highly odorous fluid has yet to be assessed. It may be that it has some sexual significance in adults, but more probably it is mainly secretory. The fact that it does, on occasion, deter aggression does not necessarily indicate its function is solely protective.

To refer again to the neanidal glands, the area surrounding the ostioles is more strongly sclerotized than the remainder of the integument and is termed a 'dorsal plate'. These plates are usually quadrate, trapeziform or elliptical in shape. In some species the plates may be conical, strongly convex oı tuberculate.

In some of the Pentatomidæ sclerotized areas are present on all segments of the abdomen of the neanide, but not more than three bear the ostioles of glands. In adult forms, although the dorsal abdominal glands are absent, vestiges of the ostioles are usually apparent, and by preparing specimens in KOH the presence of the vestigial gland sac may sometimes be revealed.

References

Balduf 1941; Beament 1946, 1947; Berlese 1914; Brindley 1930, 1934; Bueno 1906; Butler 1930; Carayon 1960; Cobben 1968; Dispons 1955; Eidmann 1924; Gadeau de Kerville 1902; Girault 1906; Grandi 1951; Gross 1901; Gupta 1961; Halasfy 1958; Heidemann 1911; Heymons 1906, 1926; Hungerford 1922, 1923; Jordan 1932; Jordan 1935; Kershaw 1908, 1909, 1910; Kirkaldy 1909a, 1909b; Kobayashi 1955; Krause 1939; Lattin 1955; Lebrun 1960; Leston and Southwood 1954; Leuckart 1835; Miller 1929a, 1929b, 1931b, 1932a, 1953b, 1955; Putshkov 1956; Putshkov and Putshkova 1956; Putshkova 1966; Readio 1926, 1927b; Sikes and Wigglesworth 1931; Slater 1951; Southwood 1949, 1956; Steer 1929; Stroyan 1954; Usinger 1946, 1947; Walton 1936; Whitfield 1929, 1933; Wigglesworth 1954; Wigglesworth and Beament 1950; Woodroffe 1963; Woodward 1952; Wygodzinsky 1944, 1947b.

Chapter 3

The Legs of Heteroptera

The legs of Heteroptera perform, apart from the act of locomotion, several other functions. For that reason it is considered appropriate to describe these functions and at the same time to bring to notice some of the remarkable modifications in the femora, tibiæ and tarsi of certain species.

In the first place, an important duty performed by the legs is that of cleaning the antennæ and rostrum from which dirt is removed by the anterior tibiæ which the bug holds together, drawing the parts to be cleaned between them. The body and wings are cleaned by being stroked by the anterior and posterior tibiæ, the former in some species having a small comb-like structure composed of short and moderately robust setæ, on the inner surface near the apex.

Female Heteroptera have been observed to rub and scrape with the posterior tibiæ the terminal segments of the abdomen. This is to remove vestiges of spermatozoal matter adhering after copulation has taken place, but it possibly could give the impression that the insect was stridulating.

As a rule the legs are used by the male to grasp the female prior to, during and sometimes after copulation. It is, however, not an invariable rule that the male continues to clasp the female at this time; in fact, when connexion has been satisfactorily made, it may release the female and orientate its body in the opposite direction, *more canum*.

Predatory species naturally use the anterior legs for seizing prey, the median legs sometimes assisting in the operation. Nevertheless, in species which are exclusively predaceous, for example, those belonging to the Reduviidæ and Nabidæ, the form of the anterior legs does not invariably conform with the types known as raptorial. In one type the anterior femora and tibiæ have teeth or spines or abundant and somewhat robust setæ, on the lower surface.

The raptorial type of leg is a characteristic of the **Emesinæ** (Reduviidæ). In the Phymatidæ as well as in some aquatic Heteroptera, namely the Belostomatidæ, Nepidæ and the Gelastocoridæ, the anterior legs are modified to form another raptorial type in which the femur and tibia may have short denticles on the inner surface and the femur may be considerably produced apically, the produced portion forming one side of a pincers (Phymatidæ), or the anterior tibia may be provided with very abundant short setæ (Belostomatidæ, Nepidæ, Gelastocoridæ). The anterior tibiæ of Enicocephalidæ are sometimes spined on the inner surface.

In some Reduviidæ and Nabidæ the anterior, and sometimes the median, tibiæ have a very useful structure known as the *fossula spongiosa* which covers the inner surface of the tibia to a varied extent.

Latreille, it would appear, was the first to note this structure, but he did not suggest what its function might be. Twenty-six years later, Dufour examined it, and gave it the name 'fossette spongieuse'. He stated 'ce corps placé au dessus du tarse et d'une forme ovalaire est charnu, pulpeux et, à sa surface inférieure paraît au microscope couvert d'un duvet excessivement court semblable à celui du velours. C'est une véritable pelote spongieuse, un organe eminément fonctionel destiné à exercer l'acte du toucher et de la préhension, et adapté aux habitudes d'insecte habituellement chasseur.'

It is clear from the foregoing that Dufour had made a fairly thorough examination of the structure in both living and dead specimens, but, unfortunately, he did not emphasize the fact that it is only after the bug is dead and the structure has desiccated and is consequently shrunken that it can be termed a 'fossette'.

In 1837, Spinola, apparently unaware of Dufour's observations, drew attention to the 'fossette' and expressed his opinion as follows: 'Les tibias de plusieurs Réduvites ont un organe particulier qui m'a paru exercer une fonction analogue à celui d'une ventouse et que j'ai nommé pour cette raison, ventouse tibiale. . . . Cette conformation a une certaine analogie avec les ventouse des sangsues qui sont étrangères à leur nutrition et qui ne le serve qu'adhérer étroitement aux corps qu'elles ont pris pour point de départ. Elle m'a paru remplir le même office dans nos Réduvites auxquelles elles prêtent un moyen facile de prendre une position verticale ou renverse, soit obliquement soit même horizontalement et de s'y maintenir en adhérant étroitment au corps solide, sans avoir besoin ni de le saisir ni de s'y cramponner'.

Ninety-two years afterwards, another observer (who had apparently overlooked the remarks of Spinola concerning the structure) concluded that it assisted the possessor to walk up a smooth surface and therefore must be a climbing organ. Later investigation, however, has demonstrated that the function of the 'fossette' is to increase the gripping capabilities of the legs.

Observation of an attack on, and the eventual overcoming of large and powerful arthropods such as Diplopoda by members of the subfamily **Ectrichodiinæ** (Reduviidæ), will confirm that without the *fossula spongiosa* the tibiæ of the attacker would not be capable of gripping the prey.

A remarkable departure from the usual methods of zoophagous Heteroptera for capturing prey is practised both in the adult and neanidal stages by the Reduviids *Amulius* Stål 1865, and *Ectinoderus* Westwood 1843, which are found in the Oriental Region and by certain members of the **Apiomerinæ** (Amyot and Serville) 1843 in the Neotropical Region. The Reduviids from the Oriental Region make use of resins (Malay-damar) produced by certain trees, in particular *Agathis alba* and *Pinus merkusii* which they smear on their anterior tibiæ. It is noteworthy that the resins selected do not harden very rapidly and thus retain their efficiency for a suitable period.

After having applied the resin, the Reduviid takes up a position on a tree-trunk with its body at an angle to the tree and often with the head directed downwards. It extends the tibiæ and waits until some small insects—usually bees of the genus *Trigona*—become entangled after the manner of a fly on fly-paper.

This striking habit of making use of resin to capture insects was apparently first observed by Uittenboogaart in Surinam in 1901. In the course of collecting other insects, he found a *Beharus lunatus* Lepeletier and Serville 1825, on a trunk of a resin-producing tree. This species belongs to the subfamily **Apiomerinæ.**

Since he was in some doubt as to whether the glutinous substance with which the tibiæ were smeared had been purposely applied or whether it had been secreted by the insect, he removed it by washing in alcohol. He records that after the removal of the substance the Reduviid, in a short space of time, found another supply and plunged the anterior tibiæ in it.

What probably gave rise to uncertainty as to the origin of the substance in the case referred to, may have arisen from the fact that the other legs had some of it on them. Attention is drawn to this because of the several species of **Apiomerinæ** examined by the writer, most of them had traces of the substance on the median and posterior legs and indeed, sometimes on the body.

Fig. 3. Legs of Heteroptera (*facing*)

1. *Scaptocoris talpa* Champion 1900. Cydnidæ-Cydninæ. Anterior tibia and tarsus.
2. *Idem*. Posterior tibia.
3. *Carcinocoris bilineatus* Distant 1903. Phymatidæ-Carcinocorinæ. Anterior femur and tibia.
4. *Agreuocoris nouhailieri* Handlirsch 1897. Phymatidæ-Macrocephalinæ. Anterior femur and tibia.
5. *Holoptilus* sp. Reduviidæ-Holoptilinæ. Anterior leg.
6. *Anoplocnemis curvipes* (Fabricius) 1781. Coreidæ-Coreinæ. Posterior leg.
7. *Gorpis papuanus* Harris 1939. Nabidæ-Gorpinæ. Anterior leg.
8. *Acocopus verrucifer* Stål 1864. Coreidæ-Merocorinæ. Posterior leg.
9. *Spalacocoris sulcatus* (Walker) 1872. Lygæidæ-Blissinæ). Anterior leg.
10. *Megenicocephalus chinai* Usinger 1946. Enicocephalidæ. Anterior leg.
11. *Ectomocoris* sp. Reduviidæ-Piratinæ. Anterior leg.
12. *Stenolæmus* sp. Reduviidæ-Emesinæ. Anterior leg.

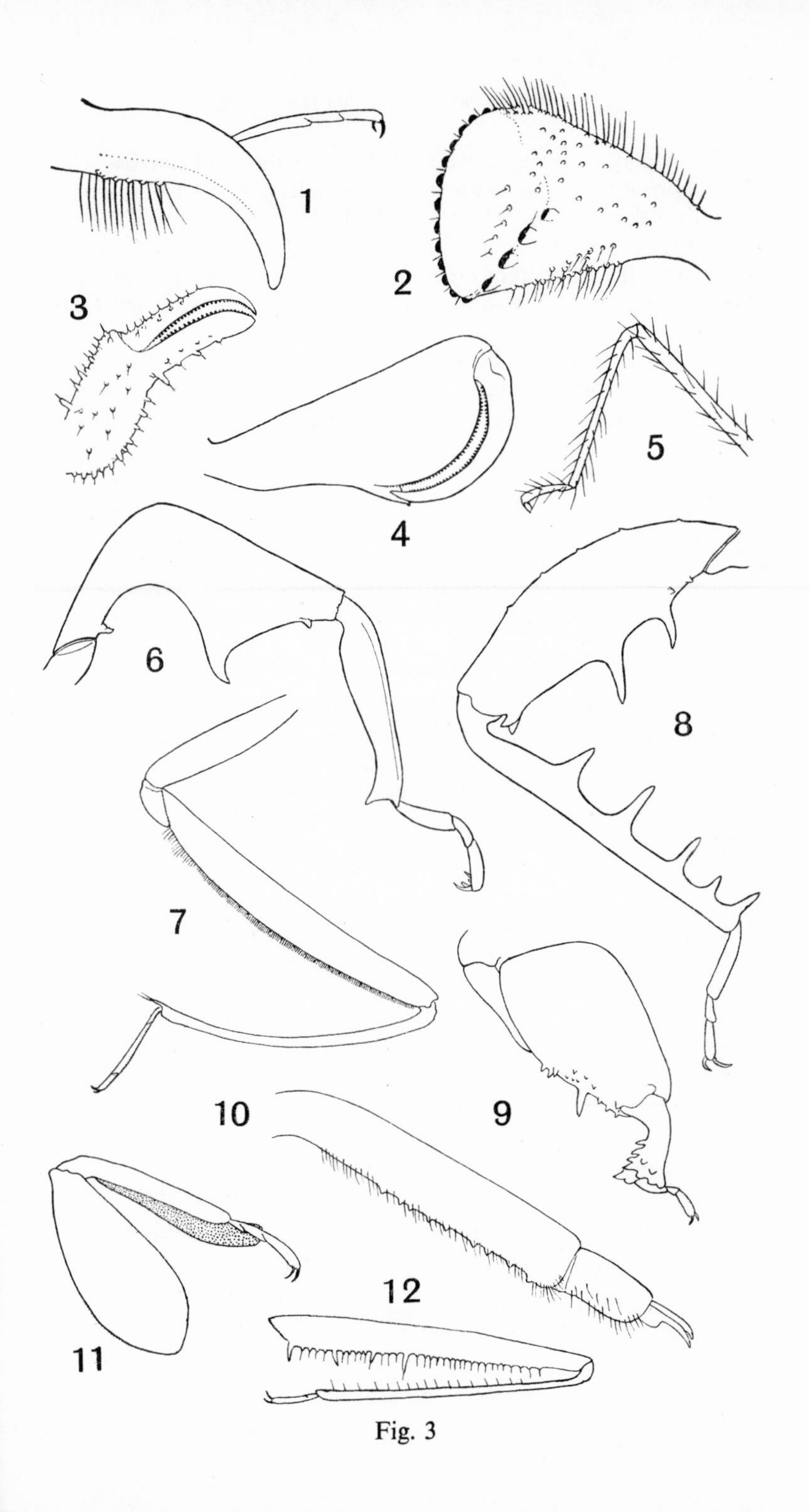

Fig. 3

The Malaysian species which belong to the **Ectinoderinæ** (Stål) 1866, appear to be able to prevent the resin from coming in to contact with the other legs and with the body. In Malaya I removed the resin from the anterior tibiæ of *Amulius malayus* Stål 1866, and was able to observe the insect renewing the resin which it did by dipping the tibiæ in a small quantity of it. The resin in this case was from *Agathis alba.*

The Malayan species do not use the anterior legs for locomotion and, furthermore, the tarsi are modified. In the South American **Apiomerinæ** the tarsi are fully developed, but the anterior legs are not used for walking.

The Ethiopian genera *Cleontes* Stål 1874, *Diaspidius* Westwood 1857 and *Rhodainiella* Schouteden 1913 (**Diaspidiinæ** Miller 1959) accumulate a sticky substance on their legs and, at times, on their bodies. The source of this substance appears to be unknown, but possibly may be a secretion from an *Acacia* species. Whether these Reduviids utilize it in the capture of prey as do the members of the **Ectinoderinæ** and **Apiomerinæ** is apparently not known.

In the first mentioned genus the anterior tarsi are lacking and in the other two genera they are reduced and lie, when not in use, in a shallow sulcus at the apex of the tibiæ.

In phytophagous Heteroptera and in those which have wholly or partly abandoned a vegetable diet, the anterior legs are, as a rule, of simple structure. The median and posterior legs, however may exhibit considerable diversity in shape and ornamentation.

Modifications to the anterior legs of *Spalacocoris sulcatus* (Walker) 1872 (Lygæidæ-Blissinæ), suggest that they perform a double function, predatory and fossorial, the femora and tibiæ being spined on the lower surface and the tibiæ also having sub-acute spines at the apex, directed, more or less, forwards.

Among the most striking differences in form between the median and posterior legs are those exhibited by many genera of the Coreidæ, for example, *Leptoglossus* Guérin 1830, *Anoplocnemis* Stål 1873,

---

Fig. 4. Legs of Heteroptera (*facing*)

1. *Chelochirus* sp. Lygæidæ-Blissinæ. Anterior leg.
2. *Diactor bilineatus* (Fabricius) 1803. Coreidæ-Meropachydinæ. Posterior tibia.
3. *Limnogeton expansum* Montandon 1896. Belostomatidæ. Anterior tibia and tarsus.
4. *Acanthocephala* sp. Coreidæ-Coreinæ. Posterior leg.
5. *Nerthra grandicollis* (Germar) 1837. Gelastocoridæ. Anterior leg.
6. *Lethocerus niloticum* (Stål) 1854. Belostomatidæ. Anterior leg.
7. *Petascelis foliaceipes* Distant 1881. Coreidæ-Coreinæ. Posterior leg.
8. *Sigara sjostedti* (Kirkaldy) 1908. Corixidæ. Anterior tibia and tarsus.
9. *Rhinocoris* sp. Reduviidæ-Harpactorinæ. Anterior leg.
10. *Sigara* sp. Corixidæ. Posterior tibia and tarsus.

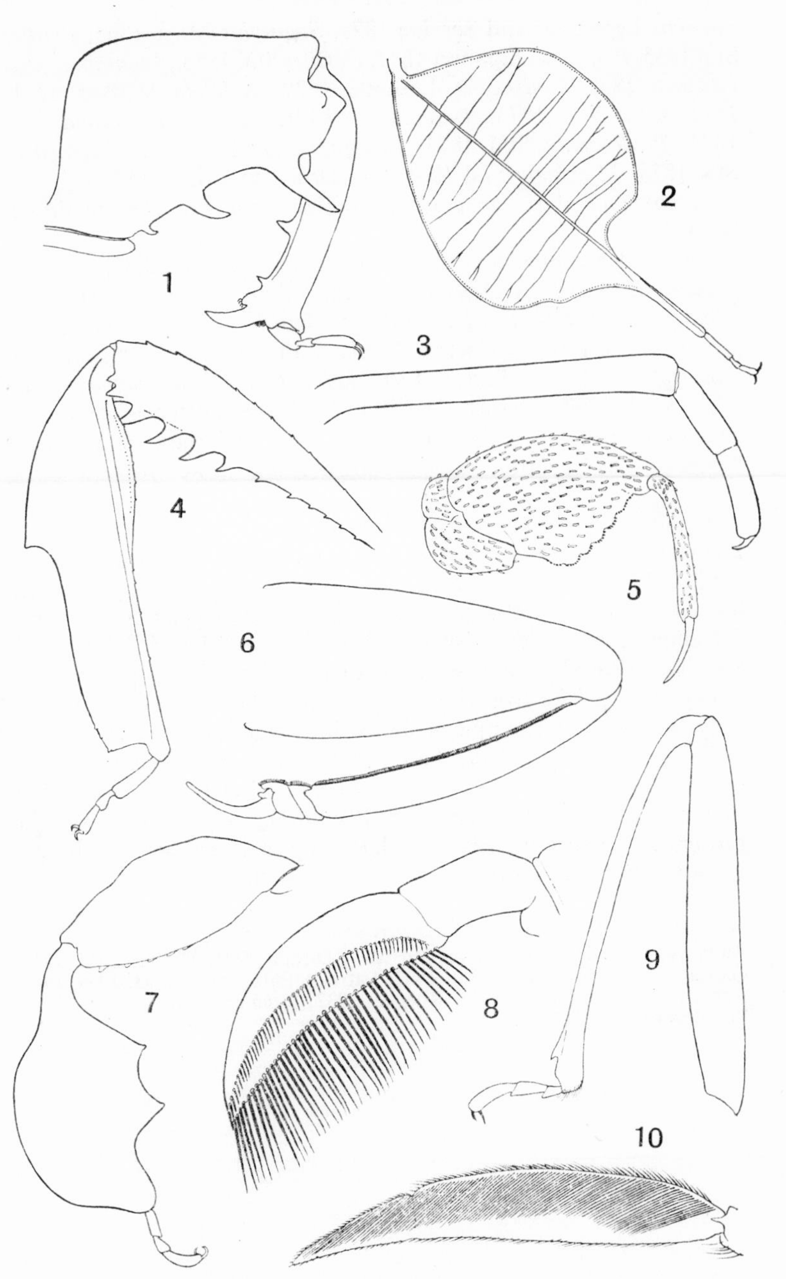

Fig. 4

Pachylis Lepeletier and Serville 1825, *Sagotylus* Mayr 1865, *Petillia* Stål 1865, *Petascelis* Signoret 1847, *Carlisis* Stål 1858, *Acanthocephala* Laporte 1832, *Anisocelis* Latreille 1829, *Sulpicia* Walker 1871, *Holcomeria* Stål 1873, *Derepteryx* White 1837, *Prionolomia* Stål 1873, *Phyllogonia* Stål 1873, *Plectrocnemia* Stål 1873, *Pternistria* Stål 1873, *Mygdonia* Stål 1865, and *Ochrochira* Stål 1883.

In some of these genera the median tibiæ are also modified, being flat.

It should be noted that these striking structural modifications are present in the males only. The modifications in the posterior legs usually take the form of an enlarged femur with the tubercles or spines on the lower surface and tibiæ with expansions or spines. In *Anisocelis*, *Diactor* Perty 1830, *Leptoglossus* and *Stenoscelidea* Westwood 1847, *Placoscelis* Stål 1867 the posterior tibiæ are strongly foliaceous. The purpose, if any, of these extravagant modifications in form and ornamentation, however, remains a mystery.

Aquatic phytophagous Heteroptera, e.g. Corixidae, have the anterior legs considerably modified and adapted for clasping food. In these legs the tarsi are fused into one segment, the pala. The other legs also perform different functions: the middle pair are adapted for clinging to some object or other; the posterior pair, of which the tibiæ have abundant setæ, are adapted for swimming.

Legs adapted for fossorial purposes are to be seen in a few genera only. An excellent example is offered by the genus *Stibaropus* Dallas 1851 (Cydnidæ), in which the anterior tibiæ are formed for use as scrapers while the posterior tibiæ serve to eject the soil which the bug excavates. These Heteroptera live among roots in the soil. The posterior tibiæ are used in stridulation by those species that have a stridulatory organ on the ventral surface of the abdomen.

References

Amyot and Serville 1843; Champion 1898; Distant 1903; Dufour 1833, 1834; Gadeau de Kerville 1902; Gillett 1932; Latreille 1807; Miller 1938, 1939, 1942; Rabaud 1923; Readio 1927; Roepke 1932; Spinola 1837; Weber 1930; Wigglesworth 1938, 1939.

Chapter 4

STRIDULATION

The ability to produce sounds by friction of one part of the body against another is possessed by many insects, including the Heteroptera. It attains, however, the highest degree of efficiency in most of the Reduviidæ, both in the adult and the more advanced neanidal stages. A stridulatory organ is also present and equally efficient in most of the Phymatidæ.

With regard to the Reduviidæ, the act of stridulation was first observed by Ray in the year 1710. Referring to *Reduvius personatus* Linnæus 1758, he wrote that it emits a sound not unlike that made by grasshoppers, by rubbing its sternum with its beak. Two hundred years later, Handlirsch described in more detail how the sounds are produced and based his description on an examination of *Coranus subapterus* De Geer 1773. In the prosternal furrow of the Reduviid he found that there were 170 transverse striæ separated from each other by a space of 0.005 mm.

In the majority of species the width and the spacing of the striæ are more or less uniform, but in some genera of the subfamily **Tribelocephalinæ** the striæ are feebly developed, widely separated from each other and few in number.

The genus *Stenolmæus* Signoret 1858 **(Emesinæ)** exhibits a somewhat different type of striated furrow. In *S. plumosus* Stål 1871 and *S. crassirostris* Stål 1871, the striæ become progressively coarser towards the posterior end of the furrow; in *S. marshalli* Distant 1903, *S. decarloi* Wygodzinsky 1947 and *S. bogdanovi* Oshanin 1870, the furrow is coarsely striate in the posterior half only.

Another type of striæ is to be seen in the Ethiopian genus *Afrodecius* Jeannel 1919 **(Tribelocephalinæ).** In this genus the sulcate part of the prosternum is somewhat arcuate and the striæ—also few in number—are of varied width.

Possession of a stridulatory furrow has been considered a characteristic of the family Reduviidæ, but increased knowledge has shown that this concept is no longer valid. Some Ethiopian genera, namely *Diaspidius* Westwood 1857, *Cleontes* Stål 1874, *Rhodainiella* Schouteden 1913 **(Diaspidiinæ),** *Phonolibes* Stål 1854 **(Phonolibinæ)** and *Aphonocoris* Miller 1950 **(Harpactorinæ),** also the Oriental genera *Amulius* Stål 1865 and *Ectinoderus* Westwood 1843 **(Ectinoderinæ)** have no striations in the furrow.

A stridulatory furrow is present in *Lophocephala* Laporte 1832, a genus of **Phonolibinæ** found in India. In the genus *Aulacogenia* Stål 1870 **(Stenocephalinæ),** the striæ in the posterior two-thirds of the furrow are coarse but in the remainder they are visible only under high magnification.

Phonolibes tricolor Bergroth 1912, an Ethiopian species, has a very long rostrum. It is straight and extends to the base of the abdomen; its apex, therefore, could not be brought into contact with striæ even if they were present.

The genus *Linshcosteus* Distant 1904, has no stridulatory furrow and the rostrum is much shorter than the head. Other genera of the **Triatominæ** in which this genus has been placed are provided with a striate prosternum.

The genus *Psophis* Stål 1863 and *Euvonymus* Distant 1904 **(Reduviinæ)** have the prosternum convex and no striæ.

There are indications that in the Oriental genus *Staliastes* Kirkaldy 1900, the striæ in the furrow appear to be degenerating and the rostrum appears to be becoming shorter. In this case also, contact is not possible between the rostrum and the striæ. In *Staliastes malayanus* Miller 1940 **(Reduviinæ),** striæ are absent and in other species of the genus, namely *rufus* Laporte 1832 and *zonatus* (Walker) 1873, the furrow is very narrow and has no striæ posteriorly. In the species in which striæ are present they are visible only under high magnification.

Campylorhyncha Stål 1874 **(Tegeinæ)** (Oriental Region), has the rostrum extended considerably beyond the posterior end of the prosternal furrow which has coarse striæ posteriorly.

In the majority of the Reduviidæ the prosternum in which the striated furrow is located, is not extended posteriorly beyond the anterior coxæ. The genera in which it is so extended are *Xenorhyncocoris* Miller 1938 **(Ectrichodiinæ)** and *Sava* Amyot and Serville 1843 **(Harpactorinæ).** It reaches well beyond the anterior coxæ. In both these genera the rostrum is very long. In *Pantoleistes* Stål 1853 **(Harpactorinæ)** the prosternum is rounded with an extremely narrow sulcus which apparently has no striæ.

A stridulatory apparatus is present in certain Pentatomidæ belonging to the subfamily **Pentatominæ.** This apparatus, which has been termed a *macula stridulatoria,* consists of an ovate, striate area on the fifth and seventh segments of the abdomen ventrally. The possessor stridulates by rubbing the posterior tibiæ—which are provided with short pegs—against it. Both sexes of *Tetyra* (Fabricius) 1803, *Pachycoris* Burmeister 1835, *Polytes* Stål 1867 have this apparatus.

A similar type of stridulatory apparatus is present in representatives of the subfamily **Tessaratominæ** (Pentatomidæ). On the first segment of the abdomen dorsally is an elliptical striate area and at the base of vein Cu of the metathoracic wings is a row of short ridges parallel to each other and in a more or less straight line.

Among the species having this type of apparatus are *Embolosterna taurus* Westwood 1837, *Hypencha opposita* Walker 1868, Malaysian species and *Eurostus validus* Dallas 1851 from China and Formosa. It is present in both sexes.

In the **Scutellerinæ** (Pentatomidæ), the genera *Sphærocoris* Burmeister 1835 and *Chiastosternum* Karsch 1898 have a stridulatory apparatus. The males of *S. testudo-grisea* (DeGeer) 1778, a common and variable Ethiopian species, have the apical margin of the pygophore produced and flattened. On this flattened portion is a strigil composed of six or seven transverse rows of robust peg-like bristles lying with the apex directed backwards. It is assumed that the apex of the scutellum acts as a plectrum.

The employment of the antennæ in the production of sounds has been referred to by several observers who have stated that the Coreid *Phyllomorpha laciniata* de Villiers 1835 is able to produce sound by rapidly vibrating its antennæ. This phenomenon has also been recorded in respect of *Centrocoris spiniger* (Fabricius) 1781 and of *Spathocera laticornis* Schilling 1829. It is doubtful if sounds perceptible to the human ear could be produced by such feeble appendages as antennæ, however rapid and sustained the vibration. It is noteworthy that vibration of antennæ may frequently be observed in neanides.

In *Rhyticoris* Costa 1863 (Coreidæ-Coreinæ) structures which I am of the opinion are the components of a stridulatory apparatus are present. The strigil is situated on the under side of the claval area basally and the plectrum on the upper side of the base, close to the costal margin of the metathoracic wing. Structures of this kind would probably produce a rustling sound similar to that caused by rubbing two pieces of rough paper together.

The faculty to stridulate is present in certain Lygæidæ, e.g. in the tribe Metrargini (Kirkaldy) 1902. Regarding these Usinger and Ashlock state, 'In the typical genus *Metrarga* the costal margin of each hemelytron is minutely crenulate, the crenulations extending for a short distance as fine striations on the under surface of the wing. Immediately opposed to this on the inner apical portion of each hind femur is an area of small tubercles that could be interpreted as a scraper. Whether the file-like costal margin serves as a strigil and the hind femur as a plectrum is a matter for conjecture because sound production has not been reported for these bugs. However, similar structures have been observed in the Pyrrhocorid *Araphe* (Lattin 1958) and in other Lygæids (Ashlock and Lattin 1960). Thus the type of stridulatory mechanism (if that is what it is proved to be) is widely scattered in Lygæid bugs and must be assumed to have arisen independently in several groups.'

One species of Piesmatidæ (Amyot and Serville) 1843, *Piesma quadratum* (Fieber) 1861, according to Schneider and Leston is able to stridulate, the apparatus being a strigil situated on vein Cu of the metathoracic wing which is scraped against a ridge on a dorsal abdominal segment more or less as in those Cydnidæ which possess the faculty.

A stridulatory mechanism has been recorded for the Aradidæ by Bergroth, the species concerned being *Artabanus excelsus* Bergroth 1892. This consists of a rastrate area on each side of the third ventral segment. Esaki and Matsuda also described the same structure when drawing up their description of *Artabanus lativentris*, but they did not mention what other part of the insect constituted the plectrum. This, as pointed out by Usinger, is provided by the posterior tibiæ which have a fine file-like ridge on the inner posterior surface. Usinger has drawn attention to four types of stridulatory mechanism in addition to that possessed by *A. excelsus*. In *Strigocoris* Usinger 1954, the posterior margin of the segment has comb-like teeth on each side and the plectrum is the sub-apical enlargement of the posterior femur. *Pictinus* Stål 1873 has a similar apparatus except that the surface of the posterior margin of the second visible segment is modified and its surface is finely file-like rather than comb-like. A fourth type is present in the males only of *Illibius* Stål 1873. The second visible segment has a wide smooth area on each side of the middle. Across the middle of this area is a minutely striate arc. Stridulation is effected by the rubbing of a single peg located on the inner surface of the posterior femur. In *Aradacantha* Costa 1865 a wide, minutely grained arc crosses the metapleuron. In this genus a ridge on the inner face of the median femora acts as a plectrum.

An apparatus presumed to have a stridulatory function, present in the males only of *Nabis flavomarginatus* Scholtz 1846, has a row of robust, curved setæ near the apex of the abdomen. By rubbing these setæ with the posterior tibiæ the bug has been said to produce sounds, although the investigator himself had been unable to hear them. This Nabid was observed to rub the apex of the abdomen first with one tibia and then with the other, a motion which may be observed often in other insects; for example flies, beetles and also other Hemiptera. The object is not to produce sounds but to remove particles of fæces or other unwanted substances.

In the case of *Nabis flavomarginatus*, the male only was concerned and the deduction was that the motion produced sounds for the purpose of attracting the female which, nevertheless, was not observed to react in the expected manner.

An unusual modification of the surface of the hemelytra which is possibly a component of a stridulatory organ, is to be seen in *Ptenidiophyes trinitatis* China 1946, of the family Dipsocoridæ. Lying obliquely to the apex of the corium is a deep longitudinal fissure, and along the inner margin of an elongate oval thickened area incorporating two veins and the area between them towards the bottom of the fissure, there is a row of ten pegs projecting from the costal side and extending to the other side. Stridulation would be effected, probably by rubbing this area with the posterior tibiæ.

Both sexes of the aquatic bug *Plea minutissima* Leach 1817, are able to produce sounds. In this species the stridulatory apparatus is formed by part of the posterior margin of the prosternum which fits into a depression in which are fine striæ. By moving the prosternum backwards and forwards the bug produces sound, but more than one bug must stridulate simultaneously for it to be audible to the human ear.

Other families of aquatic Heteroptera contain sound-producers. In the Corixidæ, Notonectidæ and Nepidæ only the males have a stridulatory apparatus, but both sexes of the Veliidæ possess one. Male Corixids are able to produce sounds of varying pitch sometimes resembling chirping and at other times the sound of a knife being rubbed against a hone. The explanation as to why the tones change is probably that the pala is armed with short teeth and the anterior femora with short spines. Both these parts of the anterior legs are rubbed against the transversely-grooved labrum, the toothed pala producing one type of note and the spined femora another.

Kirkaldy, remarking on the stridulatory organs in *Corixa*, states that they 'are so diversely formed that it is possible to distinguish the various species (in the male sex) from an examination of those organs only. This is, I believe, unique, up to the present, among the Rhynchota.'

In *Naucoris cimicoides* Linnæus 1761, Handlirsch discovered that on the dorsal surface of the abdomen of the male only there were two finely striate areas which he considered formed the plectrum of a stridulatory organ. Stridulation in this case was effected by friction between them and the apical margin of the preceding segment. Since this organ is present in the male only, he considered that the sounds produced were for the purpose of attracting the female.

Stridulation has been reported to occur in the Nepidæ but definite information regarding this is lacking. Bueno and Hungerford have stated that in a species of *Ranatra* there is present on the external surface and near the base of the anterior coxæ a low elevation, and on the inner surface of the pronotum a ridged area. Friction between these two parts is said to produce sounds.

In conclusion, it must be pointed out that, although there is an appreciable amount of information on the sound-producing capabilities of Heteroptera, a satisfactory explanation of the phenomenon is not yet forthcoming. The fact that in some instances both males and females are able to stridulate as well as neanides, means that the probability of its having a sexual significance can be ruled out. Furthermore, the act is performed in varied circumstances: during copulation and when the insect is disturbed, for example. Again it may occur for no apparent reason.

It seems clear, therefore, that stridulation is solely a nervous reaction and, with the possible exception of the Corixidæ, in the males and females of which a tympanal organ is present, there is no evidence that the sounds are perceived mutually.

References

Ashlock and Lattin 1963; Bergroth 1892; Drake and Davis 1958; Esaki and Matsuda 1951; Handlirsch 1900a, 1900b; Horváth 1894; Kirkaldy 1901; Leston 1952a, 1954a; Miller 1958; Muir 1907; Olivier 1899; Saunders 1893; Schaefer 1962; Schneider 1928; Usinger 1954.

Chapter 5

NATURAL ENEMIES OF HETEROPTERA

Heteroptera, in common with other insects, are liable to be attacked by various enemies of which the most important are hymenopterous parasites of the ova, belonging to the families Scelionidæ, Eupelmidæ and Braconidæ. I have also observed a wasp —*Polybia rhaphigaster* Saussure devouring the ova of *Eusthenes robustus* Lepeletier and Serville 1825 (Pentatomidæ-Tessaratominæ) in Malaya.

Ants, too, must be numbered among the foes from which even large and robust species with a relatively tough integument rarely escape alive; if they do, they will probably have been severely mutilated. Mammals, birds and reptiles also are enemies, but there is relatively little information regarding the extent to which Heteroptera suffer from their attacks.

It would appear that on account of the odorous fluid which most Heteroptera, both neanidal and adult, are able to secrete, attacks by vertebrates would occur infrequently, but the ability to produce an apparently disagreeable fluid does not invariably ensure protection. The victim therefore, although not killed and devoured, is likely to suffer injury sufficient to cause eventual death.

Although there is a good deal of information on the general food of insectivorous mammals, no specific instances of their eating Heteroptera have been recorded. Monkeys are known to have a mixed vegetable and animal diet which includes insects, any kind of which are probably accepted. It has not been possible to find any references of insects—particularly Heteroptera—being eaten by wild monkeys.

Experiments have been carried out in which Heteroptera have been offered to captive animals, but since the conditions of such tests are artificial, it is not reasonable to expect trustworthy results indicating the likes or dislikes of an animal.

Nocturnal species of Heteroptera are probably captured by bats, but there is no available information on the subject. Accumulations of insect remains in sheltered places to which bats go to consume the insects captured during flight have been exhaustively examined by the writer on several occasions in Southern Rhodesia but no heteropterous fragments were found among the wings, legs and other portions of Orthoptera, Coleoptera and Lepidoptera which had been rejected and were lying on the floor. It has to be recognized, nevertheless, that very many Heteroptera are small and have a delicate integument. It is possible those snapped up by a bat would be entirely devoured and assimilated.

Heteroptera are eaten by birds quite frequently. This fact has been revealed by the examination of stomach contents of birds from various parts of the world and, as will be seen from the following examples, they are all capable of secreting an odorous substance.

The stomachs of thirty-three species of birds in North America were found to contain recognizable remains of adults and neanides of Pentatomidæ including *Brachymena tenebrosa* Walker 1867, *B. quadripustulata* (Fabricius) 1775, *Chlorochroa sayi* Stål 1872, *C. uhleri* Stål 1872, *C. ligata* (Say) 1831, *Thyanta custator* (Fabricius) 1803, *Peribalus abbreviatus* (Uhler) 1872, *Euschistus inflatus* Van Duzee 1904, *E. variolanus* Palisot Beauvois 1805, *Acrosternum hilaris* (Say) 1831 and *Carpocoris remotus* Horvath 1907. In addition to these neanidal and adult forms, a large number of ova was also found.

An American bird which has been recorded as feeding on Heteroptera is the sage sparrow *Amphispiza nevadensis*. In stomachs of this species, representatives of the Pentatomidæ, Lygæidæ, Nabidæ and Miridæ were found. In India the Pentatomid *Nezara viridula* Linnæus 1758, a species which secretes a highly odorous fluid, has been found in the crops of *Dicrurus ater*, *Graculus macii*, *Oriolus melanocephalus*, *Acridotheres ginginianus* and *Francolinus vulgaris*.

To give one more example of a bird as a predator of Heteroptera, it was reported to the writer on one occasion in Malaya that the crop of a jungle fowl, *Gallus bankiva*, had been found to contain in addition to vegetable matter a large number of the Pentatomid *Tetroda histeroides* (Fabricius) 1798, a common pest of growing rice in the Oriental Region. This species has a relatively hard integument and also secretes an odorous fluid.

From these examples it is evident that the fluid does not repel insectivorous birds in every instance. Shore birds and those birds which frequent ponds and marshes occasionally capture Heteroptera, mostly aquatic kinds such as Corixidæ.

With regard to reptiles, large numbers of insects are consumed by lizards but there are few records of Heteroptera being eaten. Examination of the stomachs of certain African lizards has disclosed the remains of Pentatomidæ (*Sciocoris* sp.), Coreidæ, Lygæidæ (*Dieuches* sp.) and Reduviidæ (*Ectomocoris quadrimaculatus* Serville 1831). These Heteroptera had been devoured by the common Scincid *Mabuia striata* (Ptrs.).

There is only one record known to the writer of Heteroptera being eaten by a snake. This concerned a cobra *Naia tripudians* which he dissected. In its gut there were the remains of an unidentifiable Pentatomid as well as portions of beetles and ants.

The odorous secretion in some instances is, however, effective in repelling an enemy. Repulsion has been observed by the writer in Malaya when geckoes (*Hemidactylus frenatus*) were stalking insects on the walls and ceilings of houses at night when lights were on. At certain times of the year the Cydnid *Geotomus pygmæus* Dallas 1851, is attracted to artificial light in large numbers in buildings and is an intolerable nuisance since it is liable to fall into food and drink rendering them unfit for consumption. The odour of the fluid produced by this bug is reminiscent of that of castor oil. During hunting, a Gecko often seized a *Geotomus* but immediately rejected it. The bug is usually partially crushed, however.

Toads have been reported occasionally to eat the so-called squash bug *Anasa tristis* (De Geer) 1773 (Coreidæ) and *Euschistus fissilis* Uhler 1872 (Pentatomidæ). Aquatic Heteroptera are often preyed upon by fish (trout), in the stomachs of which remains of Gerridæ, Notonectidæ and Corixidæ have been found. Terrestrial kinds have also been discovered in the stomachs of fish, but it is probable that they were captured and swallowed when dead after having fallen accidentally into the water.

In addition to Hymenopterous parasites, Heteroptera also have other insect enemies belonging to other Orders which either kill them outright or destroy them slowly. Internal parasites belong to this category.

It has been reported that cockroaches will eat bedbugs, but experiments have shown that this is only partially true. In experiments conducted with the object of ascertaining whether cockroaches would actually prey on bedbugs, the conclusion was reached that this bug is not eaten to any great extent. Under natural conditions it is probably never an item in the varied diet of the cockroach.

The bedbug, nevertheless, is a prey of the Reduviid *Reduvius personatus* Linnæus 1758, the fairly widely distributed Palæarctic species and also probably of another Reduviid *Vesbius purpureus* Thunberg 1784 **(Harpactorinæ),** a much smaller and brightly coloured species confined to the Oriental Region. The probability that this Reduviid preys on the bedbug occurred to the writer during a prolonged stay in an internment camp in Sumatra. There, bedbugs were present in countless thousands and the only other insect found with them was *V. purpureus*.

Vesbius purpureus is often to be found beneath the floor boards of Malay-type dwellings. In this situation many spiders and cockroaches are frequently present, but *Vesbius* does not appear to prey on them. Attempts to rear this Reduviid in the laboratory, by supplying spiders and cockroaches for food failed, as did other attempts with lepidopterous larvæ. Of the several species of *Vesbius* this species appears to be the only one associated with man and, in consequence, may be distributed by his agency. An instance of this has come to notice in the discovery of *V. purpureus* in Zanzibar, whither it had probably been carried by dhow from India.

Reduviidæ are almost entirely general feeders and they also attack other Heteroptera both in the neanide and adult stages. Incidentally, cases of cannibalism in this family have frequently been reported, for example in certain species of **Triatominæ,** in which, it would appear, only neanides were involved. These inserted their mouthparts into the abdomens of newly engorged neanides and sucked up the blood contained therein. Victims of this type of feeding do not suffer any ill-effects apparently. It has been suggested by Ryckman that the term cannibalism is not an accurate descriptive term for this type of feeding. He proposed a new term 'kleptohemodeipnonism', which, being literally translated, means 'theft of a blood meal'.

Other enemies of the Heteroptera are the Lygæid *Geocoris pallens* Stål, var. *decoratus* Uhler 1877, predaceous on the false cinch bug in America, the Melyrid beetle *Collops quadrimaculatus* Fabricius, which feeds on the ova of *Blissus leucopterus* (Say), 1832, the water-bug *Notonecta undulata* Say 1832, a predator of the ova of *Belostoma* (*Zaitha*) *flumineum* Say 1832, and Diptera, among which the gall-midges (Cecidomyidæ) enemies of Tingidæ and tachinids, among which *Alophon nasalis* Bezzi, which attack Pentatomidæ and Pyrrhocoridæ.

Strepsiptera have been noted as attacking *Antestia* (Pentatomidæ), an important pest of coffee in East Africa and were responsible for rendering infertile from 20 to 80 per cent of the population of this insect. The effect of stylopization is to make the male incapable of fertilizing the female.

Other insect predators of Heteroptera include dragonflies, Asilidæ (sometimes known as robber flies) and Gryllidæ, namely *Gryllulus domesticus* Linnæus and *Gryllodes sigillatus* Walker (Orthoptera) which have been observed feeding on adults and ova of *Aphanus littoralis* Distant 1918 (Lygæidæ). Spiders are general feeders and will devour any Heteroptera which may be entrapped in their webs.

Mites are often found on Heteroptera, notably certain Reduviidæ among which *Velitra rubropicta* Amyot and Serville 1843 **(Reduviinæ)** and other species. Those most frequently selected by mites appear to pass a considerable part of their life under the loose bark of trees. It is not thought, however, that the mites are parasitic, but phoresy is the reason for their presence.

In certain subfamilies of the Pentatomidæ, namely **Scutellerinæ**

(Leach) 1815, **Pentatominæ** (Amyot and Serville), 1843, and **Acanthosomatinæ** (Stål) 1864, symbionts have been discovered in the mid-gut. Whether the presence of these organisms is deleterious or not has apparently not been decided.

Several species of Coreidæ and Lygæidæ have been found to be harbouring flagellates (Herpetomonas) in the gut and body fluid and, as is well known, pathogenic trypanosomes are found in most species of **Triatominæ** (Reduviidæ). These organisms, like the mites, are apparently innocuous. Other organisms found in the bodies of Heteroptera which are most probably injurious are nematodes.

References

Banks 1938; Conradi 1904; Corbett and Miller 1933; Cott 1934; van Deventer 1906; Frost and Macan 1948; Jourdan 1935; Kamenkova 1956; Kirkpatrick 1935–1936; Knowlton 1944; Knowlton and Nye 1946; Kunckel d'Herculais 1879; Mason and Maxwell-Lefroy 1912; Mckeown 1934; Miller and Pagden 1941; Milliken and Wadley 1922; Poisson 1930a, 1930b; Rosenkranz 1939; Ryckman 1951; Schneider 1940; Severin 1910; Taylor 1945; Wigglesworth 1936.

Chapter 6

Heteroptera Associated with Mammals and Birds

In the families Reduviidæ and Cimicidæ are to be found species associated with man, other mammals and birds. The universally distributed and best known of these is undoubtedly the bedbug *Cimex lectularius* Linnæus 1758, with its unpleasant odour and disagreeable habits. It thrives mostly in ill-kept dwelling houses and also in tropical hospitals where hygiene is not always given the attention it deserves. In concentration camps where large numbers of persons are herded together the bedbug finds conditions ideal and therefore is able to increase more or less without check on account of the fact that it is rarely possible to apply adequate control measures.

By far the most important Heteroptera associated with man and which affect his well-being are some of the **Triatominæ,** all of which are distributed in North and Central America and the West Indies with the exception of the tropicopolitan *Triatoma rubrofasciata* (De Geer) 1773, and the less abundant species *Triatoma migrans* Breddin 1903, recorded from Sumatra. There are two other species recorded from the Oriental Region, namely *T. pallidula* Miller 1941, from Malaya, and *T. novæguineæ* Miller 1958 from New Guinea.

T. rubrofasciata has been suspected of transmission of the disease Kalar Azar, also known as "oriental sore" in India: of the habits of the other three species nothing is known.

It was during the voyages of explorers about two hundred years ago that attention was first drawn to *Triatoma* species on account of their attacks on sleeping persons. It was not until 1909, however, when Chagas, working at the Instituto Oswaldo Cruz in Rio de Janeiro, discovered that these Reduviidæ were vectors of trypanosomes; he stated that their transmission took place when the bug was feeding and that the infective stages of the trypanosome were located in the salivary glands.

This was eventually disproved by Brumpt who demonstrated that the life-cycle of the trypanosome is completed in the hind gut of the bug and that the trypanosomes are present in the fæces which the bug ejects at the time of feeding or soon after.

If the person bitten, when scratching the site of puncture, which irritates on account of the injected saliva, rubs fæcal matter into it, or if fæces are introduced into an abrasion of the skin, or come into contact with the conjunctiva of the eye or mucous membrane of the mouth, entry of the trypanosomes into the blood stream takes place. Mammal reservoirs of the trypanosomes include armadillos, bats, opossums, wood-rats and squirrels.

Another Reduviid, *Apiomerus pilipes* (Fabricius) 1787 **(Apiomerinæ)** and the Cimicids *Cimex lectularius, C. hemipterus* Fabricius 1803, *C. stadleri* Horvath 1912 and *Oeciacus hirundinis* (Jenyns) 1839 and the Lygæid *Clerada apicicornis* Signoret 1863, have also been proved to be vectors of trypanosomes.

According to Usinger, although **Triatominæ** may feed on many other vertebrates, most of them have definite host preferences, while a few, namely *Mestor megistus* (Burmeister) 1835, *Triatoma infestans* (Klug) 1834, and *Rhodnius prolixus* Stål 1859, in South America, also *T. phyllosoma* (Burmeister) 1835, in Mexico, have so adapted themselves to conditions in houses occupied by human beings that they may be regarded as domestic parasites.

In some instances, the host relationships are unique; for example, the South American genus *Psammolestes* Bergroth 1911, is associated with a bird of the family Dendrocolaptidæ, *Phacelodomus rufifrons* (Wied); another, *Cavernicola pilosa* Barber 1937, is associated with bats in Panama and Brazil, *Belminus rugulosus* Stål 1859 is associated with a sloth in Costa Rica. Among other species of **Triatominæ,** *Rhodnius prolixus* is found commonly with human beings but, under natural conditions, on armadillos and *Cuniculus paca* (Linnæus).

Other records of the association of Heteroptera with mammals include that of *Polydidus armatissimus* Stål 1859, discovered by the writer in the nest of the shrew, *Crocidura coerulea* Linnæus in Malaya, and those of various species of Cimicidæ (*Cacodmus* spp.) on birds and bats, of the Reduviid *Lisarda* Stål 1859 found in sheep-folds in Southern Rhodesia, and of *Clerada nidicola* Bergroth 1914 (Lygæidæ), in the nest of an opossum in Australia.

Regarding *Lisarda* sp., when it was observed for the first time, in sheds where sheep were housed, the first impression was that it was there for the purpose of attacking sheep. Unfortunately the initial observation was not followed by further investigation. It is more probable that the Reduviid was preying on termites.

Another Lygæid was recently discovered in East Africa in the nest of a squirrel. The mammal occupying the nest was not, however, a squirrel but a member of the rat family, *Grammomys* (*Dolichurus*) *surdaster* Thomas and Wroughton 1908. The Lygæid in question is *Harmosticana garnhami* Miller 1957. *Harmostica* sp. (Lygæidæ) has been found in the nests of the squirrel *Funambulus palmarum* (Linnæus) 1766 in Southern India.

Certain Heteroptera which belong to predominately phytophagous groups have been observed to probe human skin as if they were attempting to suck blood. Recorded cases of this kind are relatively few and are confined to the Cydnidæ, Lygæidæ and Miridæ most of which, so far as is known, are plant-feeders. There are many predaceous species among the Miridæ.

In most of the instances in which phytophagous species have acted in this manner, it is probable that they were primarily in search of moisture and had been attracted by the odour of perspiration. In endeavouring to imbibe this they pierced the skin to some extent. Heteroptera behaving thus must therefore be regarded as facultative bloodsuckers and considered as taking the first step towards the adoption of carnivorous habits.

Myers, in discussing the biting of man by Hemiptera of normally phytophagous habits, has concluded that 'the vast majority of bites . . . are by insects under the influence of unusual conditions, amounting in extreme cases, as during attraction by electric light, to a complete extraction from their normal environment. The mere fact of alighting on a large vertebrate body, whether accidentally or in flight from beating or other collecting operations, brings the phytophagous insect within range of a host of stimuli, visual, thermal, tactile, olfactory, which are totally foreign to it'.

In discussing the association of insects and human beings it will not be out of place to give a few instances of the consumption by man of certain Heteroptera. As one example of this, the Pentatomid *Encosternum* (*Haplosterna*) *delagorguei* Spinola 1852, is collected by natives in Southern Rhodesia. They roast and eat the bugs either alone or mixed with other kinds of food. Other Pentatomids, namely *Aspongopus nepalensis* Westwood 1837, *A. chinensis* Dallas 1851, and *Erthesina fullo* (Thunberg) 1783, have been recorded as items in the diet of certain Asian peoples. The giant water-bug *Lethocerus indicum* (Lepeletier and Serville) 1825 is considered a delicacy of the Laos of Indo-China, and the ova of several species of Corixidæ are collected by the Mexicans; these ova are eaten by wealthy and also by poor people. The Pentatomid *Euschistes zopilotensis* Distant 1890, as well as other Heteroptera are extensively used as food in Mexico.

References

Bacot 1921; Bergroth 1914; Blanchard 1902; Bodenheimer 1951; Brumpt 1912, 1914a, 1914b; Cuthbertson 1934; Delamare Deboutteville and Paulian 1952; Kiritshenko 1949; Lent 1939; Miller 1931a, 1957; Myers 1929; Usinger 1934, 1944.

PART 2

FAMILIES OF THE HETEROPTERA

PART 2

FAMILIES OF THE HETEROPTERA

PLATASPIDÆ Dallas 1851, *Cat. Hem.* 61. (Plate I)

This family comprises large to very small, highly convex and shining insects, many of which are pests in varying degrees of importance of cultivated plants, mainly Leguminosæ (cf. *Coptosoma* Laporte 1833 and *Brachyplatys* Boisduval 1835).

The members of this family are characterized by the greatly enlarged scutellum which entirely covers the hemelytra (except the basal external area of the corium) and the metathoracic wings, leaving only the connexival segments of the abdomen partly visible. The hemelytra and metathoracic wings, when at rest, are folded under the scutellum more or less after the manner in which the metathoracic wings are folded under the elytra of Coleoptera.

Coptosomoides China 1941, *Bozius* Distant 1901 and *Tiarocoris* Vollenhoven 1863 have the labrum enlarged forming a small, membranous chamber into which the setæ are partly coiled. The setæ are extra long, for the reason, it is supposed, that members of these genera feed on mycelia of fungi which penetrate somewhat deeply into decaying wood. This supposition is based on the fact that Aradidæ also are mycetophagous and possess similar long setæ, which, however, are coiled in the head capsule.

A curious feature of the genera *Severiniella* Montrouzier 1894, *Elapheozygum* Kuhlgatz 1900 and of *Ceratocoris* White 1841, is the tuft of closely arranged setæ on the ventral surface of the seventh abdominal segment. This has not yet been critically examined, so that its function might be ascertained.

In colouration the Plataspidæ exhibit a considerable range. They may be unicolorous metallic greenish-black, brown, whitish or black with yellow or white vermiculation or spots.

Sexual dimorphism is exhibited by some genera, the head of the males bearing one or two projections anteriorly (Plate I, Figs. 1 and 4). In the genus *Ceratocoris* there are two projections either short or long, with the apex acute; in *Elapheozygum* there are also two projections but each is irregularly furcate apically, while in *Severiniella* there is one long projection with the apex bifurcate. These projections vary in their development in individuals. The head of the female has no outstanding structural features.

In *Triodocoris* Miller 1955, a dimorphic genus, the males have both the juga and vertex produced and lamellar; the head of the female is not so modified. Other dimorphic genera are *Teuthocoris*

Miller 1955 and *Glarocoris* Miller 1955, the males of which have the juga produced.

There are three genera which differ considerably from other genera of the Plataspidæ. They are *Probænops* White 1842, an Ethiopian genus which has a dull, not glabrous integument, and the pronotum gibbose anteriorly; *Bozius* Distant 1901, with a dull integument and a strongly punctate pronotum and scutellum; *Tropidotylus* Stål 1876, similarly punctate but also coarsely rugose. The last two genera are from India.

Regarding the hemelytra and metathoracic wings which, as stated previously, are folded before being placed in their resting position, in order to facilitate the folding of the former, the vein R is transversely constricted in many places; the metathoracic wing when at rest has only the basal lobe folded under.

In the genera *Apotomogonius* Montandon 1892, *Triodocoris*, *Ceratocoris* and *Gelastaspis* Kirkaldy 1902, vein 2A of the metathoracic wing has short, transverse pegs or ridges on the basal part which is somewhat thicker than the remainder of the vein. These pegs would appear to constitute the plectrum of a stridulatory apparatus. Since, however, the anal lobe of the wing is folded under when the wing is in a resting position and covers the vein, the pegs could not come into contact with another part of the body which might bear a strigil. The function of these pegs, therefore, is obscure.

Very little is known about the developmental stages of the Plataspidæ. The ova of *Brachyplatys subæneus* Westwood 1837 and also some of its neanidal instars have been described and figured. This species deposits its ova in groups with the longer axis parallel to the sub-stratum, each ovum contiguous at the base and at an angle with each other. The developmental stages of this species and also of *Coptosoma cribraria* (Fabricius) 1798, have been described by Kershaw.

The ovum of *Brachyplatys vahlii* (Fabricius) 1787 is cylindrical, somewhat flattened on the side which comes into contact with the substratum and with two rounded ridges on the upper surface. Both ends are somewhat flattened (Fig. 12, 2).

The ovum of *Probænops obtusus* Haglund 1894 (Fig. 12, 1) is of another type, being sub-ampulliform with the opercular end strongly oblique. The opposite end is feebly rounded.

The distribution of the Plataspidæ embraces mainly the Ethiopian, Oriental, Australian Regions and warm areas of the Palæarctic Region, the most strikingly dimorphic forms occur mainly in the Ethiopian Region.

References

China 1931, 1955; Carayon 1949a; Kershaw 1910; Miller 1931b.

LESTONIIDÆ China 1955, *Ann. Mag. nat. Hist.* (12), **8,** 210.

This family contains a single species, *Lestonia haustorifera* China 1955. It is allied to the Plataspidæ but differs from it in several important respects, namely, the metathoracic scent-gland ostioles are located near the middle line beneath the metathoracic acetabula and lack an auriculate peritreme; the evaporative areas, so typical of the Plataspidæ, are absent.

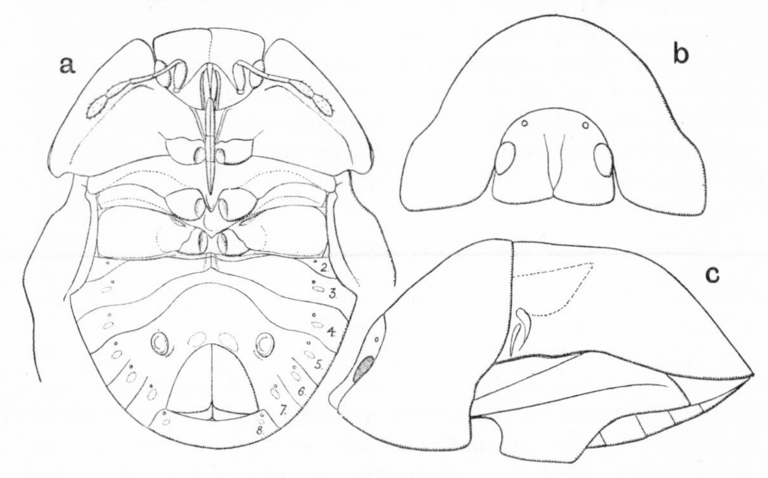

Fig. 5

Lestonia haustorifera China 1955. Lestoniidæ. (Original, China).

L. haustorifera is a small species with strongly convex dorsum and flattened ventral surface. The head anteriorly, the lateral margins of the pronotum, the costal area of the hemelytron and lateral margins of the abdomen are more or less explanate as though adapted to life closely adpressed to the surface of the food-plant.

Furthei more, a curious feature in this species is a pair of sucker-like circulai structures placed on each side of the disc between the fifth and sixth segments; these organs, depressed in the centre, are covered with very short, erect bristles similar to those on the *fossula spongiosa* of certain Reduviidæ.

Distribution, Australia, New South Wales, but the true habitat is unknown.

CYDNIDÆ (Billberg) 1820, *Enum. Ins. Billb.* 70 (Plate I)

The Cynidæ are mostly black, piceous or light brown in colour, with a shining and usually punctate integument. Although most of them are small, there are some moderately large genera, namely *Scaptocoris* Perty 1830, *Ectinopus* Dallas 1851, *Prolobodes* Amyot and Serville 1843 (Neotropical), *Cyrtomenus* Amyot and Serville 1843 (Nearctic and Neotropical) *Scoparipes* Signoret 1879 (Malaysian), *Adrisa* Amyot and Serville 1843 (Australian), *Plonisa* Signoret 1881 (Ethiopian), and *Brachypelta* Amyot and Serville 1843, which is widely distributed. Some species of *Cydnus* (Fabricius) 1803, also a widely distributed genus, are also large. There is another genus, *Sehirus* Amyot and Serville 1843, species of which are entirely violaceous or have yellowish spots. They are distributed in the Palæarctic and Nearctic Regions.

Members of the Cydnidæ live mostly in the soil and their food is mainly roots of plants and animal matter. *Geotomus pygmæus* Dallas 1851 and *Cydnus indicus* Westwood 1837, probably feed on grass roots. The former has been found among rice seedlings. The female of *Legnotus limbosus* Geoffroy 1785 has been recorded as depositing ova in a cavity in the soil and remaining with them for fifteen-twenty days, apparently without feeding.

Information regarding the developmental stages of Cydnidæ is scanty. Some work on the biology of *Sehirus bicolor* (Linnæus) 1758 has been done and the discovery of an egg-burster made. An egg-burster has also been revealed in the ovum of *Corimelæna virilis* McAttee and Malloch 1933. This is a short, highly sclerotized spine in the centre of a less sclerotized portion of the embryonic cuticle. To emerge from the ovum the neanide splits the chorion with it.

As regards structural peculiarities, the legs of some genera are formed for digging, and in *Scaptocoris talpa* Champion 1900, the posterior tibiæ have no tarsi (Fig. 3, 2); the anterior and median legs, however, are normal (Fig. 3, 1). In *Stibaropus* Dallas 1851, in which the anterior legs are fossorial, tarsi are present. The apical and lateral margins of the tylus and juga are tuberculate and have short, robust, acute spines. These parts of the head are similarly modified in *Scaptocoris* and *Chilocoris* Mayr 1864. In *Syllobus* Signoret 1879, *Lactistes* Schiödte 1848 and *Brachypelta*, the anterior margin of the head is dorso-ventrally compressed and somewhat recurved. Modifications such as these suggest a function similar to that performed by the legs.

Facultative blood-sucking has been observed in *Geotomus pygmæus*. This species is a troublesome pest from time to time on account of its invasion of dwellings to which it has been attracted by artificial light. Its presence is all the more disagreeable on account of the strongly odorous secretion from the metathoracic glands which resembles the odour of castor oil.

All the genera and species mentioned belong to the subfamily **Cydninæ.**

There are three subfamilies—namely, **Thyreocorinæ** (Amyot and

Serville) 1843, small shining insects with the scutellum extending to the apex of the abdomen, clavus visible on each side of the base of the scutellum, and with the metathoracic wing with sometimes a small hole in the anal lobe behind the second anal vein; **Cydninæ** Dallas 1851, small or moderately large insects with a flattened scutellum, head usually with distinct spines or bristles, anterior tibiæ broad and flattened; **Sehirinæ** (Amyot and Serville) 1843, with the head devoid of bristles or spines and the anterior tibiæ triangularly dilated apically.

The **Cydninæ** are widely distributed, live mainly in the soil and feed on roots, the **Thyreocorinæ** live under stones, in mammalian excreta or on plants. They are distributed in the Palæarctic Region and North and South America. No information regarding the ecology of the **Sehirinæ** is available, so far as is known. The distribution of this subfamily is world-wide.

References

Lattin 1955; Leston and Southwood 1954; Miller 1931a; Schorr 1957; Southwood and Hine 1950; Thomas 1954.

PENTATOMIDÆ (Leach) 1815,

Brewster's Edinburgh Encyclopædia, **9**, 121 (Plate I)

This is one of the most important families of the Heteroptera. It contains more than 2,500 species and a very great number is still awaiting description. Some species are of economic importance; among these are *Nezara viridula* Linnæus 1758 **(Pentatominæ),** which has a wide range of food-plants, is widely distributed and causes considerable damage to growing rice from time to time; *Solubea poecila* Dallas 1851 and *Mormidea ypsilon* Linnæus 1767 **(Pentatominæ),** are also pests of rice. Both these and *N. viridula* attack the ripening ears when they are in the 'milk' stage.

The Pentatomidæ are characterized (with exceptions) by the horizontal head, the lateral margins of which conceal the site of the insertion of the antennæ which usually have five segments, by a well-developed scutellum, odoriferous glands in both neanides and adults and tarsi with three segments. Arolia are present.

On the whole, members of the Pentatomidæ have a robust integument and exhibit an appreciable diversity in colouration which includes black, brilliant red, yellow, metallic green or blue. Sculpturation consists chiefly of puncturation and some species have spines on the head, body and legs, for example *Scotinophara* Stål 1867 **(Podopinæ),** *Aspavia* Stål 1865 **(Pentatominæ),** *Carbula* Stål 1865 **(Pentatominæ),** *Hoploxys* Dallas 1851 **(Asopinæ),** and *Leptolobus* Signoret 1835 **(Asopinæ).**

Some members of the **Tessaratominæ** are among the largest of the family and several of them have a striking appearance with the lateral pronotal angles strongly produced. Among these may be mentioned the genera *Mucanum* Amyot and Serville 1843, *Embolosterna* Stål 1870, *Pygoplatys* Dallas 1851 and *Amissus* Stål 1863, all distributed in the Malaysian sub-Region.

In this subfamily also the meso- and metasternum and the third abdominal segment are abnormal in certain genera. For example, in *Lyramorpha* Westwood 1837, *Oncomeris* Laporte 1832 and *Plisthenes* Stål 1864, the third segment of the abdomen is strongly and acutely produced medially the produced part extending to the anterior coxæ. In *Piezosternum* Amyot and Serville 1843, the mesosternum is produced and forms a wide, thick carina which extends to the anterior coxæ; the anterior part of the carina is compressed on each side.

Mucanum has the mesosternum produced forming a robust carina which extends almost to the apex of the bucculæ. The anterior end is rounded and projects outwards somewhat. The mesosternum is medially elevated so that the rostrum is displaced to one side. In *Embolosterna* the modifications are similar.

In *Siphnus* Stål 1863, the metasternum is produced and extends to the anterior coxæ. It is directed outwards anteriorly but the mesosternum is not elevated, being concave medially, so that the rostrum may lie in its normal position. The meso- and metasternum in *Hypencha* Amyot and Serville 1843, and in *Pygoplatys* are similarly modified. The rostrum in all these genera is relatively short.

Some genera are notable on account of their brilliant colouration; for example, *Mattiphus* Amyot and Serville 1843 and *Carpona* Dohrn 1863 from the Indo-Oriental Region (**Tessaratominæ**), *Callidea* Laporte 1832 and *Procilia* Stål 1864, Ethiopian genera (**Scutellerinæ**).

In the main Pentatomids are phytophagous with the exception of the **Asopinæ,** which, so far as records show, are entirely carnivorous. The best known of these are *Cantheconidea furcellata* Wolff 1801, which attacks the larvæ of the Zygænid moth *Artona catoxantha* Hampson, an important defoliator of the coconut palm in the Oriental Region; *Perilloides bioculatus* (Fabricius) 1775, which attacks the Colorado beetle in the United States of America; *Zicrona coerulea* Linnæus 1758, predaceous on *Haltica coerulea* Oliver (Coleoptera) in the Malaysian sub-region and on other insects elsewhere, since it is widely distributed, having been recorded also in many localities on the Palæarctic Region; *Oechalia consocialis* (Boisduval) 1835, a species found in Australia which preys on the larvæ of *Phalenoides glycine* (vine moth) and on *Galerucella semi-*

pullata (fig-leaf beetle). *Picromerus bidens* Linnæus 1758, a Palæarctic species has been recorded as feeding on a variety of insects, including larvæ of butterflies and moths.

According to Weber these predaceous Pentatomids are not eager to attack any but slow-moving insects which are not equipped to defend themselves. The method of attack also differs from that adopted by other predaceous Heteroptera, namely the Reduviidæ which use their anterior legs to seize their prey and then insert their stylets into a part of the body where the integument is delicate, that is, between the abdominal segments, the mouth or coxal cavities.

The **Asopinæ** rely primarily on the paralyzing effect of the saliva which they inject into their prey before seizing it with the legs. None of the **Asopinæ** has legs of the raptorial type, consequently it is not difficult to understand that they are not prone to attack lively and vigorous insects capable of defending themselves and of effecting an escape.

The genera *Cazira* Amyot and Serville 1843 and *Cecyrina* Walker 1867, both belonging to the **Asopinæ** have modified anterior legs but they are not of the raptorial type.

The habitats of Pentatomidæ are mainly the foliage and stems of plants and among plant roots. Other habitats which have come to notice are vegetable debris and flood refuse for *Eumenotes obscura* Westwood 1847 **(Eumenotinæ)**, stems of grass (*Imperata*) for *Megarrhamphus* Laporte 1832 **(Phyllocephalinæ)**, and among stems of growing rice for *Tetroda histeroides* (Fabricius) 1798 **(Phyllocephalinæ)** and *Scotinophara coarctata* (Fabricius) 1798 **(Podopinæ)**. Both *T. histeroides* and *S. coarctata* are important pests of the growing rice crop.

Pentatomidæ may also be agents in the pollination of plants. One record of this is referred to by Distant who stated that he had received information that *Cantao ocellatus* Thunberg 1784 **(Scutellerinæ)** which occurs on the 'moon tree' (*Macaranga roxburghi*) in India is a diurnal and very active species. This tree is said to depend for pollination entirely on this insect which conveys pollen to the stigma on legs, rostrum and spines.

A tendency to swarm, a factor regulated mainly by the food supply it would seem, has been observed in respect of *Nezara viridula*, *Tetroda histeroides* and *Scotinophara coarctata* congregating in rice fields: *Encosternum* (*Haplosterna*) *delagorguei* Spinola 1852 **(Tessaratominæ)** appears in large numbers at certain periods of the year and is collected and eaten by natives.

Swarming and periodic migration occurs with *Eurygaster integriceps* Puton 1881 **(Scutellerinæ)** an important pest of wheat in Asia Minor. In this species development proceeds in the lowlands, then the adults migrate to upland regions where they hibernate: they return to the plains for reproductive purposes.

The female Pentatomids usually deposit their ova in groups on part of the host-plant, but occasionally on some nearby object. They attach them to the substratum with a glutinous substance and when the act of oviposition is completed the female departs.

There are, however, certain Pentatomids, the female of which have been stated to remain on or near the group of ova until the neanides have hatched. The purposes of this unusual behaviour lacks logical explanation.

Behaviour of this kind is often interpreted as being a manifestation of maternal solicitude. Such an interpretation, however, is surely an example of the loose manner in which some observers endow organisms which are very low in the scale of development with the sentiments of animals much higher in the scale.

This peculiar habit has certainly nothing to do with the incubation of the ova and furthermore, it can have no protective value against adverse climatic conditions or against potential enemies, hymenopterous parasites and the like. A possible explanation seems to be that it is a persistence of the gregarious habit which may have been more usual formerly, a habit which persists in varying degrees in a few species.

The assumption that maternal solicitude is exhibited when a female remains with the egg-mass, is based on an observation by Dodd, in respect of *Tectocoris lineola* (Fabricius) 1781 var. *banksi* Donovan 1805 **(Scutellerinæ).** This observer recorded that *Tectocoris* clasped the group of ova which it had deposited on a plant stem and moved its position only when some object approached. An analogous reaction may be observed in Delphacidæ (Homoptera), Reduviidæ and also Buprestidæ (Coleoptera) at rest on a grass stem or a twig.

A further objection may be advanced as to the so-called protective attitude and maternal solicitude of female bugs. This objection is, that, if the female in most, if not all reported instances has not been under close examination without intermission during the entire period of incubation, which is not likely to be much less than ten days, it cannot be affirmed accurately that the bug has not moved at all from the spot during that period.

Dodd stated, however, "it is absolutely certain that 'broody bugs' remain foodless during the whole period of three weeks or more of sitting; they occupy the same position always and various investigators have failed to reveal any puncture in the twigs in front of them".

That a bug might remain without food for a period of weeks cannot be denied, but the question must be posed: which group of ova would have the 'protection' of the female in view of the likelihood that more than one group of ova is produced, the second batch following before the hatching of the first?

Some species of Pentatomidæ are very prolific and, so far as is known, deposit their ova in groups of varying numbers, with the longer axis vertical tothe substratum: but, at least, one exception in which the longer axis is in a horizontal position, is to be seen in the deposition of ova by *Megymenum brevicorne* (Fabricius) 1787 **(Dinidorinæ).** Other species of this genus and probably of allied genera deposit their ova in this fashion.

The deposition of a constant number of ova sometimes occurs. It has been observed in the large species *Pycanum ponderosum* Stål 1854 **(Tessaratominæ).** This species deposits fourteen ova always in the order three-four-four-three. The same number and arrangement have been recorded for *Pentatoma rufipes* Linnæus 1758 **(Pentatominæ).**

The ova are usually placed on the leaves or branches of the host-plant or on some object in the vicinity. Eclosion is effected by the embryo pushing off the operculum; before it can do this it has to loosen it. It is assisted in this process by the 'egg-burster', a highly sclerotized part of the embryonic cuticle. This piece of apparatus usually has the shape of an elongate, flattened rod expanded at one or both ends with a short, conical projection on the outer surface (Fig. 1), at one end. The manner in which this apparatus is used in removing the operculum is set forth in detail in the chapter on development.

The neanides of Pentatomidæ, which pass through five instars during the course of development do not resemble the adults very closely except as regards the shape of the head and legs. In colour they differ entirely from the adults.

An appreciable number of ova has been described and, in some cases, figured. From knowledge at present available it is possible to state that the ova of representatives of the **Pentatominæ, Asopinæ, Phyllocephalinæ, Dinidorinæ** and **Podopinæ** are mostly cylindrical with or without chorionic processes and those of the **Scutellerinæ** and **Tessaratominæ** spherical or ovate; chorionic processes are sometimes present in the ova of the former.

Obviously a vast number of ova remains to be described, including those of subfamilies not mentioned here. Some examples of Pentatomid ova are given in Fig. 12.

Both adults and neanides of the Pentatomidæ are able to secrete an odorous volatile fluid, the adult from the metathoracic glands with the ostiole and the evaporative area situated between the meso- and metapleura, the neanides from glands located under the dorsum of the abdomen.

The part of the integument in which the ostioles of the abdominal glands are situated is highly sclerotized and usually has the form of an elliptical or trapezoidal plate, these shapes varying, however, according to the instar of the neanide. These ostiole-bearing plates, known as 'dorsal plates' are on or between the third, fourth and fifth segments. In some neanides other segments have sclerotized plates but they do not bear ostioles. This may indicate, however, that formerly more than three segments bore gland ostioles.

The odour of the secretion produced by the glands is not always of a kind which is unpleasant to human beings, but this, of course, is an individual matter. Its action, too, is not always repellent to other organisms. It is not always possible to state to what other odour it can be likened but various odours have been suggested, for example, those of almonds, pineapple, castor oil or rotting apples.

The fluid, on being secreted from the gland, spreads over the area around the ostiole or further, according to how copious the discharge is. It then volatilizes rapidly, and should it come into contact with human skin it usually leaves a stain resembling that produced by iodine.

It is of interest that the ostiole of adults in certain genera can be seen only with difficulty since the peritreme is lacking and it is barely larger than the surrounding puncturation.

There are thirteen subfamilies of Pentatomidæ which may be characterized as follows: **Asopinæ** (Amyot and Serville) 1843. Predaceous; head horizontal; rostrum basally thickened; bucculæ short, enabling a more extensive lateral movement of the rostrum. In some species the anterior tibiæ have a short, acute spine on the lower surface. Distributed in all zoogeographical regions (Fig. 6). **Tessaratominæ** Stål 1865. Mostly large or very large; spiracles of second abdominal segment always exposed; metathoracic wing cell with a short hamus; connexivum dilated. Distribution: world-wide. **Eumenotinæ** Esaki 1922; body more or less compressed; antennæ with four segments; tarsi with two segments; scutellum extending to fourth abdominal segment, much longer than wide. Distribution: Oriental Region. **Cyrtocorinæ** Distant 1880; small, obscurely coloured; habitus somewhat bizarre; scutellum large with a strong spine or projection medially. Distribution: Neotropical Region. **Dinidorinæ** Stål 1870; mostly moderately large, black or brown with a yellow or red connexivum which is often spotted; antennæ with five segments, the apical segment somewhat compressed in some species; rostrum short, not extending beyond the mesosternum and often much shorter; bucculæ closed posteriorly; hemelytral membrane reticulate. Distribution: world-wide. **Phyllocephalinæ** (Amyot and Serville) 1843; moderately large, elongate or elliptical in outline; head acute anteriorly; juga in some species separated; pronotum transverse with the postero-lateral angles produced in some genera; rostrum mostly short, extending just beyond anterior margin of the prosternum; longer in some genera; basal segment and a large part of segment two concealed within bucculæ. Distribution: Ethiopian, Oriental and Australian Regions. **Pentatominæ** (Amyot and Serville) 1843; basal rostral segment longer than bucculæ; scutellum mostly longer than wide; tibiæ sulcate on outer surface. Distribution: world-wide. **Scutellerinæ** (Leach) 1815; scutellum covering entirely or partly the abdomen and wings. Distribution: Ethiopian, Oriental and Palæarctic Regions. **Podopinæ** (Amyot and Serville) 1843; scutellum longer than wide, often covering entire abdomen and hemelytra except corium; eyes prominent, pedunculate; antennæ more or less clavate. Distribution: Oriental, Australian, Ethiopian, Palæarctic and Nearctic Regions. **Serbaninæ** Leston 1953; bucculæ obsolescent, shorter than basal rostral segment; scutellum short, as long as wide. Distribution: Malaysian sub-region. **Acanthosomatinæ** (Stål) 1864; tarsi with

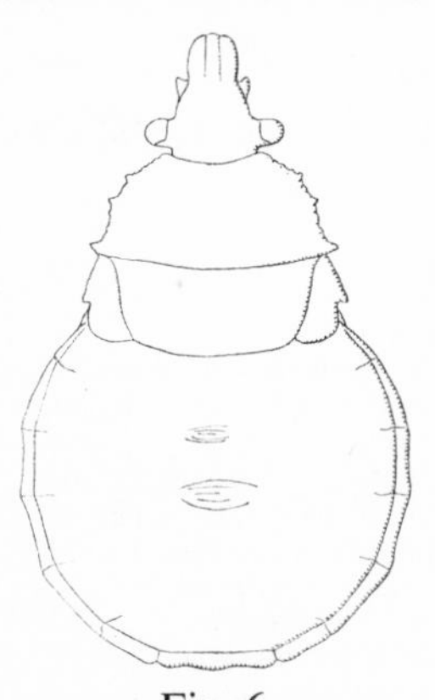

Fig. 6

Tahitocoris cheesmanae Yang. 1935 (Pentatomidæ-Asopinæ).

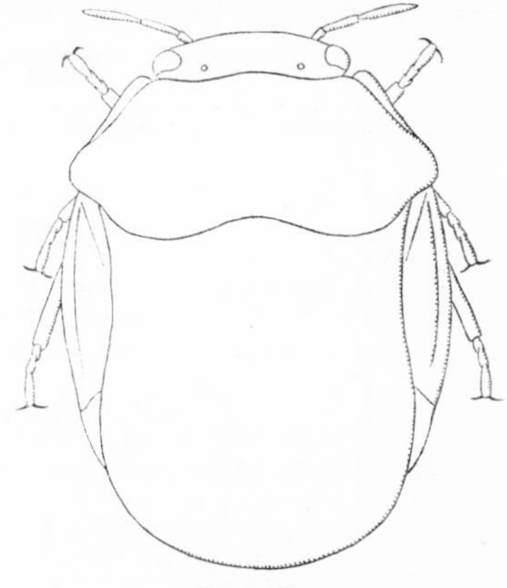

Fig. 7

Aphylum syntheticum Bergroth 1906 (Aphylidæ).

two segments; third abdominal segment spinously produced midventrally; mesosternum with a median laminiform carina; seventh abdominal segment in males in some genera spinously produced; produced portion of third abdominal segment overlaps the posterior end of the sternal carina, both ends thus in contact, being obliquely truncate; sternal carina extending sometimes almost to apex of head; antennæ with four or five (mostly) segments. Distribution: Africa, India, Java, Philippine Islands, Australia, Palæarctic Region, Mexico. *Acanthosoma hæmorroidalis* Linnæus 1758 and *Clinocoris* (*Elasmostethus*) *griseus* Linnæus 1758 have been recorded feeding on carrion. Cannibalism has also been known to occur, but the members of this subfamily are mostly phytophagous. **Canopinæ** (Amyot and Serville) 1843 previously placed in the Plataspidæ by McAtee and Malloch and in the Cydnidæ by China and Miller; small, black, shining; antennæ with five segments, the second segment very short; anterior pronotal margin rounded with lateral margins. Distribution: Neotropical Region. **Megaridinæ** McAtee and Malloch 1928; also previously placed in the Plataspidæ by McAtee and Malloch and in the Cydnidæ by China and Miller; minute blackish insects; tarsi with two segments; hemelytral membrane without venation. Distribution: Neotropical Region.

References

China and Miller 1955, 1959; Halaszfy 1958; Hibrauoi 1930; Jones and Coppel 1963; Jordan 1958; Kobayashi 1951, 1953, 1954, 1955; Leston 1955; McAtee and Malloch 1928; Miller 1929a; Pendergrast 1958, 1963; Putschkova 1959; Vodjani 1964; Weber 1930; Yuksel 1958.

Plate I (*facing*)

Plataspidæ, Cydnidæ, Pentatomidæ, Urostylidæ, Phlœidæ.

1 and 2. *Ceratocoris cephalicus* Montandon 1899. Plataspidæ.
3. *Libyaspis wahlbergi* (Stål) 1863. Plataspidæ.
4. *Severinella cameroni* Distant 1902. Plataspidæ.
5. *Cantharodes jaspideus* Fairmaire 1858. Plataspidæ.
6. *Plonisa tartarea* Stal 1853. Cydnidæ-Cydninæ.
7. *Deroplax nigrofasciatus* Distant 1898. Pentatomidæ-Scutellerinæ.
8. *Poecilocoris nigricollis* Horvath 1912. Pentatomidæ-Scutellerinæ.
9. *Solenostethium liligerum* Thunberg 1783. Pentatomidæ-Scutellerinæ.
10. *Chrysophara excellens* (Burmeister) 1834. Pentatomidæ-Scutellerinæ.
11. *Eumecopus longicornis* Dallas 1851. Pentatomidæ-Pentatominæ.
12. *Cinxia limbata* (Fabricius) 1803. Pentatomidæ-Pentatominæ.
13. *Carpona imperialis* Dohrn 1863. Pentatomidæ-Tessaratominæ.
14. *Caura marginata* Distant 1880. Pentatomidæ-Pentatominæ.
15. *Vulsirea variegata* Drury 1773. Pentatomidæ-Pentatominæ.
16. *Chalcocoris anchorago* Drury 1782. Pentatomidæ-Pentatominæ.
17. *Pygoplatys lancifer* Walker 1861. Pentatomidæ-Tessaratominæ.
18. *Leptolobus eburneatus* Karsch 1892. Pentatomidæ-Asopinæ.
19. *Edessa cornuta* Burmeister 1885. Pentatomidæ-Pentatominæ.
20. *Runibia decorata* Dallas 1851. Pentatomidæ-Pentatominæ.
21. *Urusa crassa* Walker 1868. Pentatomidæ-Dinidorinæ.
22. *Alcaeus varicornis* (Westwood) 1842. Pentatomidæ-Pentatominæ.
23. *Megymenum quadratum* Vollenhoven 1868. Pentatomidæ-Dinidorinæ.
24. *Anaxandra* sp. Pentatomidæ-Acanthosomatinæ.
25. *Urostylis farinaria* Distant 1901. Urostylidæ-Urostylinæ.
26. *Urostylis striicornis* Scott 1896. Urostylidæ-Urostylinæ.
27. *Phloea corticata* Drury 1773. Phlœidæ.

Plate I

APHYLIDÆ (Bergroth) 1906, *Zool. Anz.*, **29** (Fig. 7)

Small, dull-coloured, strongly convex insects with piceous puncturation and brownish reddish suffusion. There is only one genus known up to the present: this is *Aphylum*, of which there are two species, *syntheticum* Bergroth 1906 and *bergrothi* Schouteden 1906.

The Aphylidæ are similar in habitus to the Plataspidæ but are not glabrous as most of that family. They may be also distinguished by the strongly punctate pronotum and scutellum and by the shape of the pronotum, the lateral angles of which are lobately produced posteriorly and the postero-lateral margin strongly angulately incised.

The meso- and metanotum are visible from above, the former as a lobe in the angulate incision of the posterior margin and the latter as a segment immediately behind the lateral pronotal angles.

Furthermore, the corium (the exposed part) is highly sclerotized and resembles the scutellum in structure. The costal margin is widely separated at the base from the body margin.

The antennæ have five segments, the tarsi three, and the femora are somewhat compressed laterally with the apical part of the lower surface somewhat sulcate.

Aphylum has been recorded so far only from Australia and nothing is apparently known of its ecology.

References Bergroth 1906; China 1955; Schouteden 1906.

UROSTYLIDÆ Dallas 1851, *Cat. Hem.* 313 (Plate I)

This small family is distributed mainly in the Eastern Palæarctic Region, the Malay Peninsula and Australasia. It was formerly considered to be a subfamily of the Pentatomidæ.

Two species, *Urochela distincta* Distant 1900 and *U. falloui* Reuter 1888, have been reported as pests, the former on account of its being a nuisance when it appears in large swarms and the latter because of its destructive activities to pear trees, grape-vines and other plants in China. Apart from this information nothing appears to be known about the habits and developmental stages of species belonging to this family.

Most of the Urostylidæ are somewhat small insects, mainly pale greenish or greenish-brown in colour and have a relatively delicate integument.

The ova of *Urostylis farinaria* Distant 1901 (Fig. 12, 10) and of *Urolabida khasiana* Distant 1887 (Fig. 12, 11), which were obtained by dissection by the writer, are oval with three closely sited long filamentous chorionic processes at the upper end. There is no operculum, therefore the embryo has to split the chorion at the time of eclosion. It is possible that an egg-burster may be present.

The presence of such processes, which are similar to those on the ova of certain Miridæ (cf. *Helopeltis* Signoret 1858) suggests that the ova are inserted by the female into a relatively soft substance such as decaying vegetable matter or the young shoots of plants. On the other hand, the female may embed them in the soil.

The Urostylidæ are divided into two subfamilies: **Urostylinæ** Dallas 1851, the members of which have the corium irregularly punctate on the disc, a short rostrum not extending to the median coxæ and the metathoracic wings with normal pentatomoid venation. **Sailerioliinæ** China and Slater 1956, minute, with the corium unpunctured on disc between the distinct row of punctures along RM and the row along the claval suture; rostrum long, extending to the fourth abdominal segment. The diagnosis of this subfamily is based on a single specimen from Borneo.

References

China and Slater 1956; Miller 1953; Yamada 1914, 1915; Yang 1936.

PHLŒIDÆ (Amyot and Serville) 1843, *Hist. nat. Hem.* xxiv, 115 (Plate I)

This family contains two genera only. They are characterized by the strong, foliaceous expansion of the head, pronotum and connexivum, antennæ composed of three segments which are concealed by the pronotum, and by the dorso-ventrally compressed habitus. The rostrum is long and extends to about the middle of the ventral surface of the abdomen.

Phlœidæ are found mainly on the trunks of trees, the bark of which it is probable they may resemble according to the kind of tree on which they rest.

The ova which are cylindrical and white are deposited by the female in fissures in the bark. Apparently the ovum has no operculum, neither is an egg-burster present.

A good many apparently superficial observations have been made on the habits of the Phlœidæ and according to them, the females remain with the ova and when the neanides hatch they cluster around her and do not leave until they are more mature.

The two genera *Phloea* Lepeletier and Serville 1825 and *Phloeophana* Kirkaldy 1908 are Neotropical.

References

Brien 1930; Lent and Jurberg 1966; Leston 1953; Perez 1904.

COREIDÆ Leach 1815, *Brewster's Edinburgh Encyclopædia* **9,** 121 (Figs. 8 and 9) (Plate II)

This is a large family comprising many diverse forms. It reaches its highest state of development in the tropics, where it is represented by many genera.

The Coreidæ have the antennæ situated on the upper sides of the head; they are composed of four segments. The pronotum is commonly trapeziform but may have the angles spinose or foliaceous. Trichobothria are present on the ventral abdominal segments. The ostioles of the metathoracic glands are very distinct.

Genera with pronotal modifications include members of the subfamily **Coreinæ** (Stål) 1867, namely, *Derepteryx* White 1829, *Holcomeria* Stål 1873, *Prionolomia* Stål 1873, *Phyllogonia* Stål 1873, *Holopterna* Stål 1873, *Evagrius* Distant 1901, *Petillia* Stål 1865, *Dalade*r Amyot and Serville 1843, and *Acanthocephala* Laporte 1832. Both the lateral margins of the abdomen and pronotum are very strongly foliaceous and spinose in *Phyllomorpha* Laporte 1832, *Pephricus* Amyot and Serville 1843 and *Craspedum* Amyot and Serville 1843. These genera also of the **Coreinæ** are the most bizarre representatives of the family.

Generally speaking the hemelytra are complete, but brachypterous forms also occur. Of these may be mentioned *Typhlocolpura* Breddin 1890, *Lygæopharus* Stål 1870 **(Coreinæ),** *Psotilnus* Stål 1859, *Micrælytra* Laporte 1832, *Dulichius* Stål 1865 **(Alydinæ),** *Chorosoma* Curtis 1830 and *Jadera* Stål 1860 **(Rhopalinæ).**

One apterous or micropterous genus known so far, is the Australian *Agriopocoris* Miller 1953, of Aradid-like appearance, which was discovered among leaf debris on the floor of the forests and under bark **(Agriopocorinæ)**; recently macropterous forms have been discovered.

Ocelli are present in the Coreidæ. The scutellum is small and always shorter than the abdomen. Some tropical species exhibit striking modifications of the posterior femora and tibiæ, the former being very greatly enlarged, sometimes spinose and tuberculate and the latter may have either foliaceous expansions or spines.

In colour the Coreidæ are mostly brown or stramineous. There are, however, some genera, for example, *Anisocelis* Latreille 1829 **(Coreinæ),** which has both body and legs brightly coloured. Others, namely *Spathophora* Amyot and Serville 1843 **(Meropachydinæ),** *Machtima* Amyot and Serville 1843, *Pachylis* Lepeletier and Serville 1825 and *Golema* Amyot and Serville 1843 and also some species of *Mictis* Leach 1814 **(Coreinæ)** are brown with a red or yellow pattern.

Metallic green species, in which the colour may be due to the presence of scales or to the sculpture of the integument are found in the genera *Mictis*, *Petalops* Amyot and Serville 1843, *Phthia* Stål 1862 and *Sphictyrtus* Stål 1859 **(Coreinæ).** Dull black or piceous genera include *Hygia* Uhler 1861 and *Typhlocolpura* **(Coreinæ).**

Coreidæ possess 'stink' or repugnatorial glands, the fluid from which is usually pungent. One is often made aware of the presence of Coreidæ by their odour, although the insects themselves may not be visible; for example, in grassy areas and rice-fields where species of *Leptocorisa* Latreille 1825 **(Alydinæ)** are present in large numbers.

The adults and neanides, in certain cases, are able to project the fluid for a short distance, but as a rule, when secreted, it spreads over the greater part of the body before volatilizing.

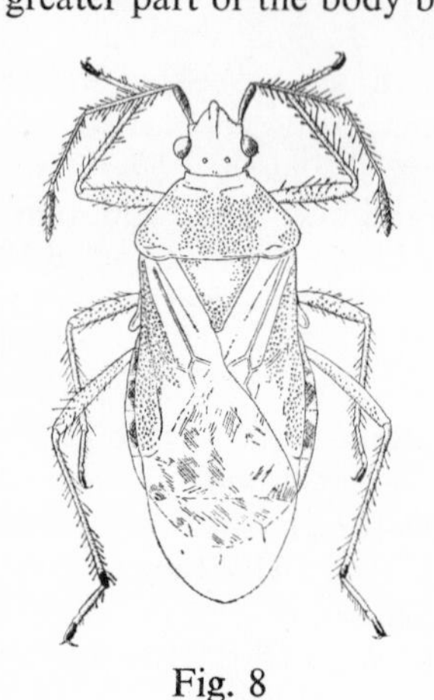

Fig. 8

Corizus latus Jakowleff 1882 (Coreidæ-Rhopalinæ).

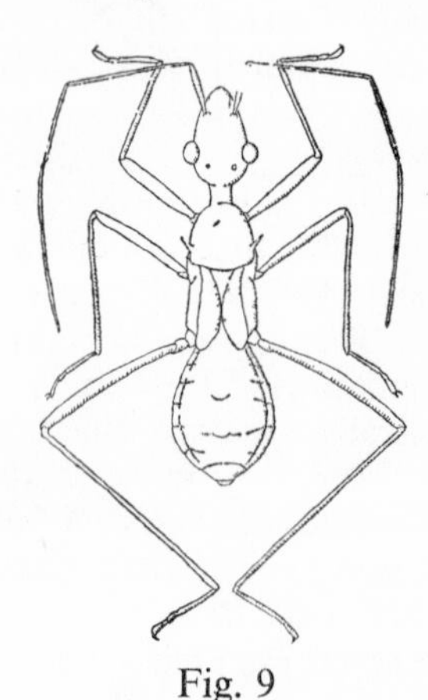

Fig. 9

Dulichius inflatus Kirby (Coreidæ-Alydinæ).

All Coreidæ are apparently phytophagous. They attack various parts but mainly the younger shoots and the leaves. The result of their attacks is the wilting and, if the attack be prolonged, the demise of the plant. This is caused by the saliva which is injected at the time of feeding.

Many Coreidæ attack cultivated plants but they may not necessarily be pests of economic importance. Some genera, however, are definitely in the category of pests. Among these are species of *Leptocorisa* which cause important damage to rice grains when they are in the 'milk' stage; *Theraptus* Stål 1859 and *Amblypelta* Stål 1873, reported to cause excessive nut-fall in coconut palms, and *Leptoglossus* Guérin 1836, a pest of many cultivated plants, including Cucurbitaceæ **(Coreinæ).**

The genera *Physomerus* Burmeister 1835, *Anoplocnemis* Stål 1873

(Coreinæ) and *Riptortus* Stål **1859 (Alydinæ),** include species which are occasional but less serious pests.

Gregarious tendencies have been observed in *Physomerus*, *Petascelis* Signoret 1847 **(Coreinæ)** and others. When such congregations occur it is not uncommon to see all stages of the bugs at one time.

Coreidæ oviposit on various parts of the host-plant, the ova being arranged in groups of varying sizes, in chains; or they are deposited singly. The arrangement may vary according to the conformation of the substratum.

Sometimes the female secretes a wax-like substance in granular or powder form which covers the exposed part of the ovum (cf. *Mictis tenebrosa* (Fabricius) 1787) and allied genera **(Coreinæ).**

An exceptional mode of oviposition has been recorded: in this the female places the ova on the male and they remain in this position until eclosion takes place.

This mode has been noticed in *Phyllomorpha laciniata* de Villiers 1835, a Palæarctic species found, according to some authors, on foliage and under stones and leaves. It also occurs on **Gramineæ** and herbaceous plants.

It should be remarked, however, that both the male and the female have been observed to have ova on their backs. This would suggest that the presence of ova on both sexes was fortuitous, the ova having fallen from a female that happened to be near.

Costa Lima records that a male specimen of *Plunentis porosus* Stål 1859 **(Coreinæ),** examined by him, had several ova adhering to the ventral surface of the abdomen and among them a recently emerged neanide without doubt belonging to the same species.

I have observed that in captivity females of the Reduviids *Bagauda lucifugus* McAttee and Malloch 1926 **(Emesinæ),** and of *Rhaphidosoma circumvagans* Stål 1855 **(Rhaphidosomatinæ),** Malayan and Ethiopian species respectively will place their ova on the legs or body of other members in the cage. This occurrence has also been observed by Brown in the Solomon Islands. He records that, in captivity, *Amblypelta* has occasionally oviposited on the bodies of other individuals.

Closer and repeated observation on ovipositing females would, no doubt, throw more light on these aberrant modes of oviposition.

Coreidæ are mostly active insects and fly readily when disturbed. So far as is known, in general, they do not take up a particular attitude when resting. One example, however, of a definite resting attitude is exhibited by *Hypselopus annulicornis* Stål 1855 **(Alydinæ).** A specimen shown on Plate II, Fig. 13, mounted in the resting position was received from Dr. E. Burtt from Tanganyika Territory.

With the exception of the ova of *Anoplocnemis*, *Derepteryx*,

Euagona Dallas 1852 **(Coreinæ)** (Fig. 12, 19), and of some other large genera which have a comparatively tough chorion, the ova of many genera have a somewhat delicate chorion. An operculum is generally present and, so far as is known, an egg-burster. Those egg-bursters which have been examined are in the form of an arcuate rod with the centre enlarged (Fig. 1).

Investigations up to the present time show that the ova of Coreidæ are of several widely different types, the most common being cylindrical, truncate at each end or ovate with the side which is in contact with the substratum flattened. Some ova have on the flattened side a short pedicel.

Peculiar forms are exhibited by ova of species of *Choerommatus* Amyot and Serville 1843 (Fig. 12, 21), and of *Catorhintha mendica* Stål 1870 **(Coreinæ)** (Fig. 12, 22). These are oblong cubical with the sides somewhat concave.

Other aberrant forms may be seen in the ova of certain **Rhopalinæ.** For example, those of *Corizus rubricosus* Bolivar 1879 (Fig. 12, 15) are ovate with the opercular end oblique and with a short process on the operculum; those of *Myrmus miriformis* (Fallèn) 1807 (Fig. 12, 17) and *Liorhyssus hyalinus* (Fabricius) 1794 (Fig. 12, 14), similar, but with a short process on the chorion below the opercular suture as well as on the operculum. These processes are apparently analogous to the chorionic processes on ova of Pentatomidæ and function similarly.

The six subfamilies of the Coreidæ are: **Agriopocorinæ** Miller 1953; micropterous, macropterous or apterous; posterior acetabula not excised; bucculæ long, extending beyond the insertion of the antennæ; spiracles located close to the margin of the abdomen, those on segments two and three marginal and visible from above. Distribution: Australia. **Rhopalinæ** (Amyot and Serville) 1843; medium to small insects with, in some cases, the interveinal areas of the corium hyaline; metapleural gland ostioles absent; if present, then situated between posterior acetabula. Distribution: world-wide. **Alydinæ** (Amyot and Serville) 1843; narrow, elongate insects; bucculæ short; head usually wider than pronotum, the posterior angles of which are sometimes spinously produced; posterior femora in males often thickened and spinose on lower surface. Medium-sized insects of world-wide distribution living on grasses and Leguminosæ. **Meropachydinæ** (Stål) 1867; head small, much shorter and narrower than pronotum; posterior femora strongly incrassate; posterior tibiæ flattened and spined; scutellum horizontally produced and with a vertical elevation. Distribution: Neotropical Region. **Pseudophlœinæ** (Stål) 1867; small, usually setose insects; head without a median sulcus; posterior femora tuberculate and with

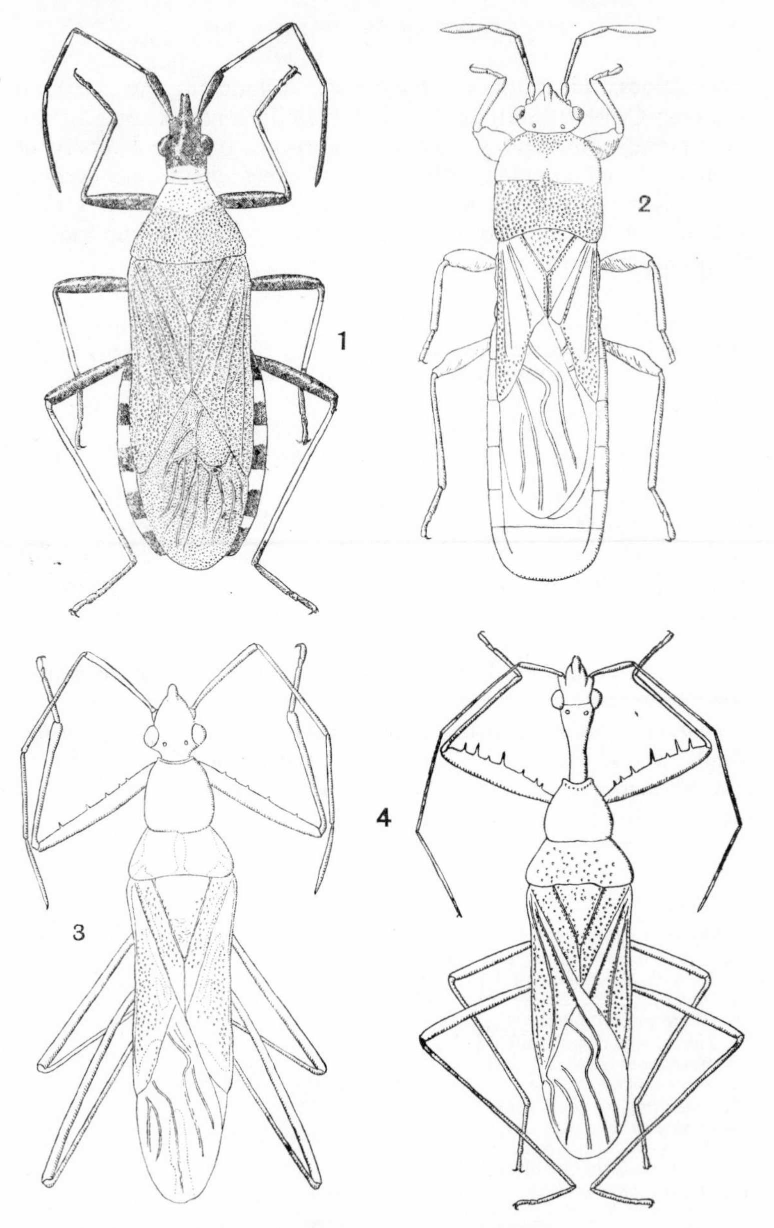

Fig. 10. Stenocephalidæ, Lygæidæ

1. *Dicranocephalus agilis* (Scopoli) 1763 (Stenocephalidæ).
2. *Macropes excavatus* Distant 1901 (Lygæidæ-Blissinæ).
3. *Narbo fasciatus* Distant 1901 (Lygæidæ-Megalonotinæ).
4. *Myodocha intermedia* Distant 1882 (Lygæidæ-Megalonotinæ).

long spines. Distribution: Palæarctic, Ethiopian and Oriental Regions; **Coreinæ** (Stål) 1867; this subfamily contains some of the largest and conspicuous genera, comprising a great diversity of forms classified into many tribes; head narrow; antennæ slender, but some segments may be expanded; head in front of eyes with a median culcus; numerous species are of economic importance. Distribution: world-wide.

References

Bolivar 1894; Brown 1958a, 1958b, 1959b, 1959c; Costa Lima 1940; Miller 1929b, 1953c; Pagden 1928; Poisson 1930; Putshkova 1955, 1957; Reuter 1909; Saunders 1893; Stroyan 1954; Szent-Ivany and Catley 1960; Stys 1964.

STENOCEPHALIDÆ Dallas 1852, *Cat. Hemipt. Brit. Mus.* **2,** 480 (Fig. 10)

This small family contains medium-sized insects which were formerly placed in the tribe Stenocephalini of the Coreidæ. The principal characters are the regularly punctate but not areolate corium and clavus, the tarsi with three segments, the membrane of the hemelytron with five or six sinuate veins, the juga acuminate and contiguous at base in front of tylus, the basal antennal segment strongly thickened and much thicker than remaining segments, densely, setosely pubescent.

Dicranocephalus agilis (Scopoli) 1763 (Fig. 10, 1) and *D. medius* (Mulsant and Rey) 1870, both Palæarctic species feed on spurge (Euphorbiaceæ), and it is presumed that species living in other areas also feed on this and allied plants.

Plate II (*facing*)

Coreidæ

1. *Pachylis pharaonis* (Herbst.) 1784.
2. *Petascelis remipes* Signoret 1874.
3. *Prionolomia malaya* Stål 1865.
4. *Acanthocephala latipes* Drury 1782.
5. *Plectrocnemia lobata* Haglund 1895.
6. *Derepteryx chinai* Miller 1931.
7. *Hormambogaster expansus* Karsch 1892.
8. *Spartocera pantomima* (Distant) 1901.
9. *Machtima mexicana* Stål 1870.
10. *Leptoglossus zonatus* Dallas 1852.
11. *Anisocelis flavolineata* Blanchard 1859.
12. *Petalops distinctus* Montandon 1895.
13. *Hypselopus annulicornis* Stål 1855.
14. *Holymenia histrio* (Fabricius) 1803.
15. *Pephricus livingstoni* Westwood 1857.
16. *Leptocorisa acuta* Thunberg 1783.
17. *Leptocoris rufomarginata* (Fabricius) 1794.
18. *Dulichius inflatus* Kirby 1891.
19. *Corizus latus* Jakowleff 1882.

Note: Nos. 1-15 Coreinæ; 16 and 18 Alydinæ; 17 and 19 Rhopalinæ.

The ova are deposited on the stems of the food-plant. Adults hibernate beneath bark or in plant detritus.

Distribution: Palæarctic, Oriental, Sonoran and Neotropical Regions (Galapagos Islands).

HYOCEPHALIDÆ Bergroth 1906, *Zool. Anz.* **29,** 649 (Fig. 11)

This family contains one genus, *Hyocephalus* Bergroth 1906. It also has one species, *aprugnus* Bergroth 1906, which was originally described from a single macropterous female from South Australia. Bergroth considered it to be a coreid with a modified lygæid-like head and the hemelytra of an aradid.

The characters are: head elongate, porrect; ocelli present; paraclypei short; anteclypeus prolonged; bucculæ large; antennæ with four segments; pronotum trapezoidal with raised lateral margins;

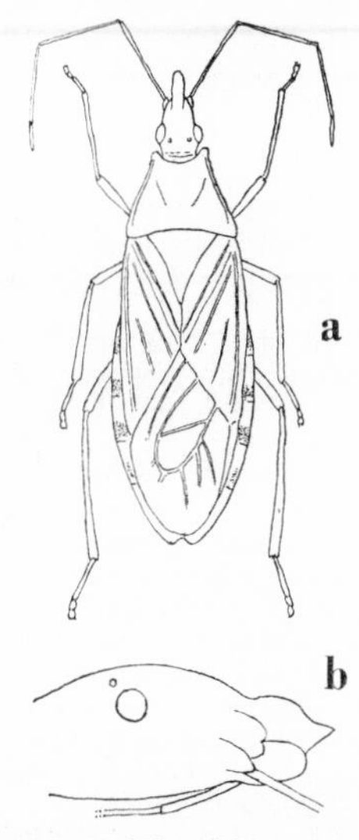

Fig. 11

Hyocephalus aprugnus Bergroth 1906. A. Whole insect, dorsal view (after Stys) B. Head, lateral view (after Bergroth) (Hyocephalidæ).

scutellum longer than claval commissure; hemelytral membrane with four veins enclosing three large cells and then becoming branched; tarsi with three segments; membranous arolia present; ovipositor laciniate.

H. aprugnus is a reddish-brown species with parts of the head and prothorax lighter coloured. The membrane of the hemelytra is blackish-grey with a reddish or brown shading.

It is probably a very old genus and a relic of an extinct group which branched from the Lygæidæ.

Nothing, so far, is known of the ecology of this family, but the shape of the ovipositor suggests that the ova when deposited, are inserted into the substratum.

Its distribution is probably confined to the south-eastern part of Australia.

Reference Stys 1964.

LYGÆIDÆ (Schilling) 1829, *Beit. z. Ent.* **1,** 37 (Plate 3) (Fig. 10)

A moderately large family of mostly sombre coloured insects, but also containing some species which have bright colours, mainly red and yellow. The Lygæidæ are characterized by having the membrane of the hemelytra with four or five sinuate veins, three-segmented tarsi and the anterior femora sometimes incrassate and spined. The antennæ are inserted below the lower margin of the eyes. Alary polymorphism occurs.

Fig. 12 (*facing*)

Ova of Plataspidæ, Pentatomidæ, Phlœidæ, Urostylidæ, Coreidæ, Lygæidæ, Pyrrhocoridæ, Velocipedidæ, Piesmatidæ

1. *Probaenops obtusus* Haglund 1895. 1·60 mm. Plataspidæ.
2. *Brachyplatys vahlii* (Fabricius) 1787. 0·80 mm. Plataspidæ.
3. *Megarrhamphus truncatus* Westwood 1837. 1·60 mm. Pentatomidæ-Phyllocephalinæ.
4. *Halyomorpha viridescens* (Walker) 1867. 1·50 mm. Pentatomidæ-Pentatominæ.
5. *Dryptocephala brullei* Laporte 1832. 1·50 mm. Pentatomidæ-Podopinæ.
6. *Mecidea* sp. 0·70 mm. Pentatomidæ-Pentatominæ.
7. *Aspongopus* sp. 1·80 mm. Pentatomidæ-Dinidorinæ.
8. *Serbana borneensis* Distant. 2·00 mm. Pentatomidæ-Serbaninæ.
9. *Phloea corticata* Drury 1773. 2·60 mm. Phlœidæ.
10. *Urostylis farinaria* Distant 1901. 0·80 mm. Urostylidæ.
11. *Urolabida khasiana* Distant 1887. 1·00 mm. Urostylidæ.
12. *Spartocera fusca* Thunberg 1783. 2·00 mm. Coreidæ-Coreinæ.
13. *Sciophyrus* sp. 1·60 mm. Coreidæ-Coreinæ.
14. *Liorhyssus hyalinus* (Fabricius) 1794. 0·70 mm. Coreidæ-Rhopalinæ.
15. *Corizus rubricosus* Bolivar 1897. 1·00 mm. Coreidæ-Coreinæ.
16. *Pephricus livingstoni* Westwood 1857. 1·70 mm. Coreidæ-Coreinæ.
17. *Myrmus miriformis* (Fallén) 1807. 1·10 mm. Coreidæ-Rhopalinæ.
18. *Sephena limbata* Stål 1862. 2·40 mm. Coreidæ-Coreinæ.
19. *Euagona diana* Dallas 1852. 3·00 mm. Coreidæ-Coreinæ.
20. *Acanthosoma brevirostris* Stål 1873. 1·00 mm. Coreidæ-Acanthosominæ.
21. *Choerommatus* sp. 1·00 mm. Coreidæ-Coreinæ.
22. *Catorhintha mendica* Stål 1870. 1·20 mm. Coreidæ-Coreinæ.
23. *Dieuches* sp. 1·50 mm. Lygæidæ-Megalonotinæ.
24. *Aphanus littoralis* Distant 1918. 1·20 mm. Lygæidæ-Megalonotinæ.
25. *Nysius inconspicuus* Distant 1903. 1·40 mm. Lygæidæ-Lygæinæ.
26. *Antilochus nigripes* Burmeister. Pyrrhocoridæ.
27. *Scotomedes alienus* (Distant) 1904. 1·70 mm. Velocipedidæ.
28. *Piesma quadratum* Fieber 1844. 0·64 mm. Piesmatidæ.

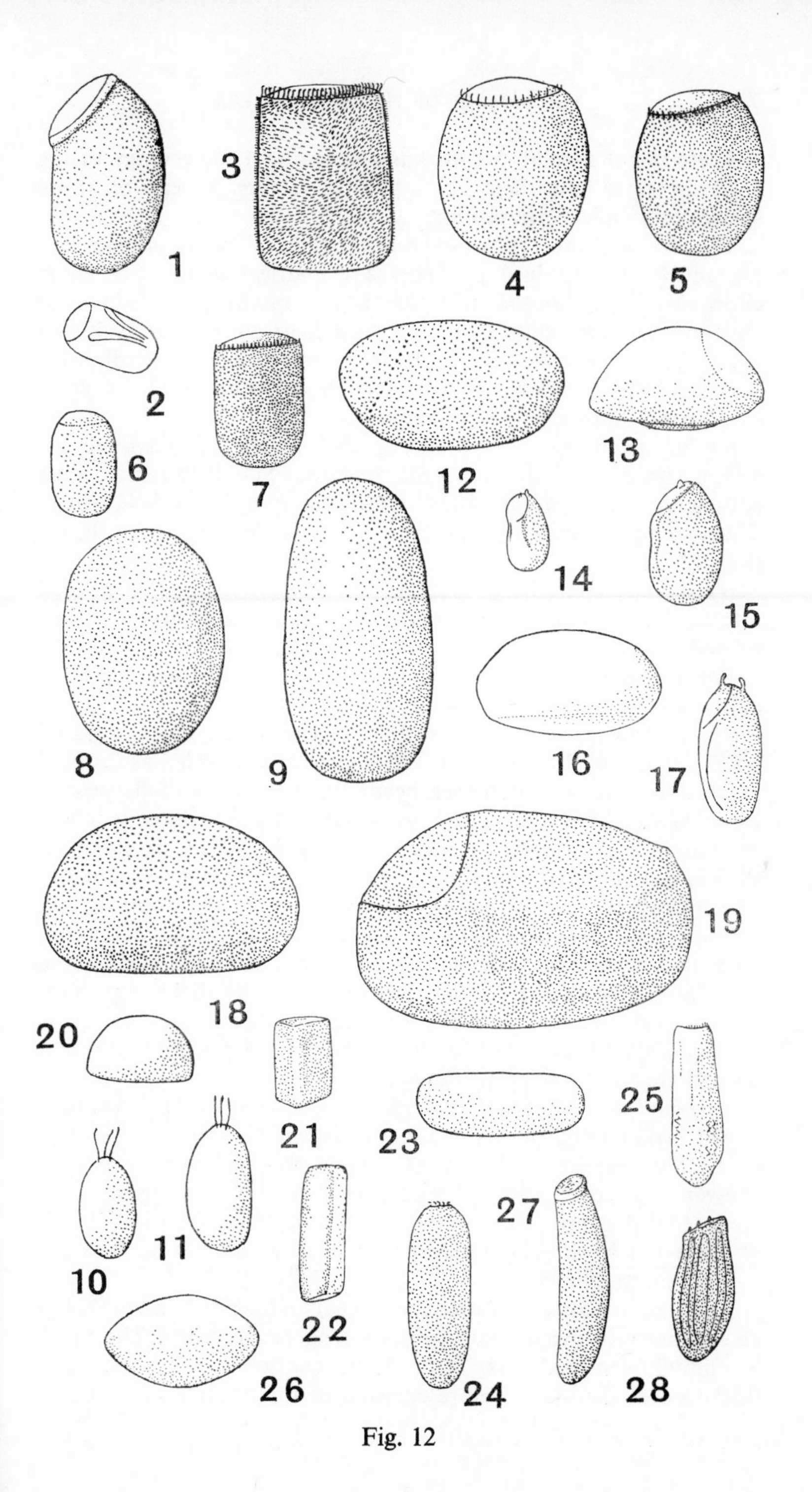

Fig. 12

There is not a great deal of knowledge about the ecology of the Lygæidæ or about their developmental stages, except of those species economically important.

The ova of a few species have been studied. *Nysius inconspicuus* Distant 1903 **(Lygæinæ)** produces a cylindrical ovum with short chorionic processes and with the base depressed; the ovum of *Aphanus littoralis* Distant 1918 **(Megalonotinæ)** is similar, but it appears to be deposited in a different manner without an adhesive; it is regularly cylindrical and rounded at each end, the micropylar end having short, recurved processes.

Nysius inconspicuus has been captured on tea (*Thea sinensis* L.) and, according to Usinger, *Nysius pulchellus* Stål 1839 occurs most commonly on *Euphorbia hirta*, *Portulaca*, *Pemphis* and *Vernonia*. The ova are deposited somewhat loosely in composite heads (Fig. 12, 25).

Nysius caledoniæ Distant 1920, occirs on *Emelia*, a plant introduced into the island of Guam, and it has also been collected on the closely related composite *Erigon* in the Philippine Islands.

Blissus leucopterus (Say) 1832 **(Blissinæ),** is an important pest of cultivated **Gramineæ.** Another species referred to by Usinger, *Pachybrachius nigriceps* (Dallas) 1852 **(Megalonotinæ)** was captured on *Tournefortia*, *Styphelia*, *Euphorbia hirta* and recorded as depositing ova on the top of flower heads among the small flowers of Heliotrope. They were placed crosswise and rather loosely in the open, but were fastened together and to the calyx. They are elongate, slightly curved and about 1 mm. in length by 0.35 mm. in diameter. The micropylar end is thicker, sub-truncate and the opposite end tapered and rounded. The chorion is sculptured, exhibiting prominent hexagonal reticulations anteriorly. Near the middle, the reticulations become inconspicuous or appear as slight rugosities. The posterior end is perfectly smooth and the entire surface is shining white. There are five small, but distinct processes forming a small ring at the micropylar end.

Aphanus littoralis has been recorded as feeding on freshly harvested, stored groundnuts, on sesamum, *Carthemus tinctorius*, millet and on *Solanum nigrum*. It is not yet clear in what manner the ova are deposited by the female but, it would seem, from the few observations made, that they are placed in loose groups in the soil, or in the case of those females attacking products in storehouses, among dust and rubbish on the floors.

A species of *Cænocoris* (up to the present undescribed and probably new) deposits its ova in a mass with the longer axis vertical. The ovum is ampulliform and has 14–17 long, chorionic processes, each thickened at the apex. Before eclosion of the neanide the apex of each process is curved inwards. There is no operculum, the ovum being ruptured at its apex by the emerging neanide. An egg-burster is present. This is a relatively large, more highly sclerotized portion of the embryonic cuticle, with a still more strongly sclerotized transverse ridge. The total length of the ovum, 2.50 mm.

The ova on which this information is based were deposited on a leaf of *Thevetia peruviana* K. Schum, an ornamental plant introduced into Malaya. The natural foodplant of *Cænocoris* is unknown, and it is possible that its presence on *Thevetia* and its oviposition thereon were fortuitous. The number of ova in the mass was approximately 70.

Apart from those quoted, oviposition methods of Lygæidæ are hardly known at all, but it is not unlikely that some species insert their ova into plant tissues or in crevices in bark or other material.

Some species of *Oxycarenus* Fieber 1836 **(Oxycareninæ)** are pests of the cotton plant on the seeds of which they feed.

A large number of members of the **Megalonotinæ** which contains nearly half the known species of the Lygæidæ, feed on plant seeds. A considerable number of species, on account of their strongly incrassate and usually spined anterior femora are suspected to be predaceous in habit.

The anterior legs of some **Blissinæ,** for example, of *Spalacocoris sulcatus* Walker 1872, and of a species of *Chelochirus* Spinola 1839 (Fig. 4, 1) are strongly spined, which indicates that these Lygæids also have a carnivorous diet. The anterior tibiæ of the former have forwardly directed spines at the apex which suggests that they may be used for excavation purposes also.

Geocoris liolestes Hesse 1947 **(Geocorinæ)** has been recorded as a predator of the red scale (Coccidæ) of Citrus in South Africa and a species of *Geocoris* Fallèn 1814 has been reported as preying also on Coccidæ as well as the larvæ of *Papilio demodocus* (Lepidoptera) which feed on Citrus.

Geocoris species have also been found feeding on various plant bugs.

I have once found a species of *Dieuches* Dohrn 1860 **(Megalonotinæ)** feeding on the excreta of *Hyrax* sp. (Mammalia) and *Oncopeltus jucundus* Dallas 1852 **(Lygæinæ)** feeding on millipedes crushed on a road in Southern Rhodesia.

Habitats of Lygæidæ include plants and under stones among leaf debris. One myrmecophilous species has been described, namely *Neoblissus parasitaster* Bergroth 1903 **(Blissinæ)**, which lives in the nest of *Solenopsis geminata* Fabricius in Brazil. It is thought to feed on the food stores of the ant.

Some species are nocturnal and are attracted to artificial light. Among these may be mentioned *Dieuches* sp.

The nest of the Indian Grey Squirrel in Ceylon has provided a habitat for a species of *Harmostica* Bergroth 1918, and *Harmosticana garnhami* Miller 1957, was found in the abandoned nest of a squirrel taken over by the 'tree rat' *Grammomys* (*Dolichurus*) *surdaster* Thomas and Wroughton 1908. Whether these Lygæids were actually preying on the mammals in question or were there for the purpose of feeding on their excreta, is not certain.

The ovum of *H. garnhami* is elongate, cylindrical and narrowly rounded at each end. Odoriferous glands are present in the neanides and, so far as observation goes, are situated on the third and fourth abdominal segments.

Alate species of the tribe Metrargini (Kirkaldy) 1902 have been reported to be gregarious, very many congregating together at the base of the leaves of *Freycinetia* in the Hawaiian Islands. Neanides also are found mingling with the adults. Alate individuals are also to be found amongst dead leaves and fragments of fern frond. Incidentally, the odour of the glandular fluid of *Metrarga* B. White 1878 species is said to be disgusting, and when a group of these bugs is disturbed, it taints the surrounding air.

There are eighteen subfamilies: **Megalonotinæ** Slater 1957; anterior femora thick and often spined; the third ventral suture not extending to lateral margin; spiracles usually ventral, but the basal three segments or some of them often with the spiracles dorsal. Widely distributed. **Geocorinæ** (Stål) 1862; in this subfamily the head is very wide, the pronotum wider than long, with a median transverse sulcus; scutellum longer than wide; hemelytra convex and punctate. Distributed in all zoogeographical regions except Mascarene. **Blissinæ** (Stål) 1862; mostly small or very small, somewhat elongate insects; all spiracles dorsal except on segment 7 where they are ventral; head usually narrower than posterior margin of pronotum; anterior tibiæ sometimes modified for fossorial or raptorial purposes. A large subfamily in which the species are mainly phytophagous; some species possibly carnivorous; one known species myrmecophilous. Distributed in all zoogeographical regions except Mascarene. **Cyminæ** (Stål) 1862; small insects with head, pronotum and scutellum punctate and the internal veins of the membrane not connected at the base by a transverse vein; hemelytra wider than abdomen with the corium distinctly punctate, its costal margin being dilated. Distributed in all zoogeographical regions. **Ischnorhynchinæ** (Stål), were originally placed by Stål (1872) as a tribe of **Cyminæ,** but raised to family rank by Usinger and Ashlock 1959 (*Proc. Hawaiian ent. Soc.* **17,** 100) following suggestions by Slater and Hurlbutt 1957 (wing), Ashlock 1957 (male genitalia), and Southwood 1956 (egg). They are distinguished by the distinct pronotal calli and by the arrangement of the hemelytral punctures in regular lines instead of the dense arrangement in **Cyminæ. Lygæinæ** (Stål) 1862; contains some of the larger members of the subfamily; many species brightly coloured; hemelytra not wider than abdomen and the costal margin of the corium not dilated; head, pronotum, scutellum and corium usually impunctate; the two internal veins of the membrane generally joined at the base by a transverse vein; usually phytophagous; one species recorded as feeding on carrion; some species of economic importance. Distributed in all zoogeographical regions. **Orsillinæ** (Stål). The **Orsillinæ** were originally placed as a tribe of **Lygæinæ** but were raised to subfamily rank by Scudder (1957) without reason, but the change was supported by Usinger and Ashlock 1959, following suggestions by Slater and Hurlbutt 1957 (wing venation), Ashlock 1957 (male genitalia) and Southwood 1956 (ova). They are distinguished from the **Lygæinæ** by having the apical margin of the corium sinuate instead of straight, as well as by the characters studied by the workers listed above. **Oxycareninæ** (Stål) 1862; small insects with the anterior femora moderately incrassate and armed with a spine; the lateral margins of the pronotum not laminate and the lateral margins of the corium, which is much wider than the abdomen, laminate; mostly dull-coloured and somewhat setose phytophagous insects, some of which are of economic importance. Distributed in all zoogeographical regions. **Bledionotinæ** Reuter 1878; very small, somewhat delicate, pallid insects with the head, including the eyes, wider than the pronotum; scutellum as long as wide; pronotum longer than wide with a transverse sulcus near middle; head, pronotum, scutellum and corium strongly punctate; femora moderately thick. Distribution, West Indies. **Malcinæ** (Stål) 1866; very small insects with relatively long antennæ, the basal and apical segments of which are thickened; ocelli more or less contiguous; connexival segments 5–7 with laminate expansion; head, pronotum and corium strongly punctate. Distributed in the Oriental Region. **Lipostemmatinæ** (Berg) 1879; very small, oblong elliptical, somewhat depressed insects, densely pilose with a moderately wide head; eyes large with large facets; membrane with three or four veins; ocelli absent; pronotum trapezoidal. Distribution, Neotropical Region. **Henestarinæ** (Douglas and Scott) 1865; small insects, oblong in outline with large, pedunculate eyes; basal segments of antennæ thicker than remaining segments, extending beyond apex of head; all femora moderately incrassate; pronotum as wide as long; head, body and corium remotely pilose; spiracles on segments 2, 6 and 7 ventral, on segments 3–5 dorsal. Distributed in the Palæarctic Region. **Pachygronthinæ** (Stål) 1865; small to moderately large, elongate, pale-coloured insects; antennæ with basal segment longer than head and pronotum together; anterior femora incrassate and with spines on lower surface; rostrum not extending to, or only a little beyond anterior coxæ; segments 5 and 6 of abdomen midventrally wide; anterior tibiæ shorter than femora; basal segment of tarsi longer than segments 2 and 3 together. Distribution, Nearctic, Palæarctic, Sonoran, Ethiopian and Australian Regions. **Heterogastrinæ** (Stål) 1872; small, somewhat hirsute species, with the pronotum wider than long and with the lateral margins dorso-ventrally compressed, straight and feebly constricted; anterior femora not incrassate and spinose on lower surface. Distributed in all zoogeographical regions. **Chauliopinæ** Breddin 1907; spiracles on dorsal side of connexivum; corium shorter than membrane; connexival segments 5–7 elevated; eyes stylate; antennophore with a distinct projection in front of eye. Distributed in the Ethiopian and Oriental Regions. **Artheneinæ** (Stål) 1872; very small insects with the pronotum as long as wide and with the lateral margins dorso-ventrally compressed, straight; basal antennal segment not extending to apex of head; scutellum with a deep depression; anterior femora moderately incrassate; anterior tibiæ shorter than femora. Distributed in Palæarctic Region. **Phasmosomatinæ** (Kiritshenko) 1938; elongate species; ocelli absent; head acuminate; antennæ and legs long and slender; brachypterous. Distribution Transcaucasia. **Henicocorinæ** Woodward 1968. Broadly elliptical in outline. Head porrect without trichobothria; ocelli present, widely separated. Bucculæ short, conjoined immediately behind base of labium. Antennæ and legs of moderate length. Fore coxal cavities closed; fore femora not strongly incrassate, without ventral spines; tibiæ not sulcate; tarsi three-segmented. Sides of pronotum and costal margin of corium not laminate. All abdominal spiracles ventral, those of segment 11 in membranes near anterolateral angles of sternum. Abdominal sterna 3 and 4 with a single median trichobothrium on each side, without lateral trichobothria; sternum 5 without trichobothria; sterna 6 and 7 each with two lateral on each side. Scars of abdominal scent gland openings between terga IV–V and V–VI.

References

Bergroth 1903; Corby 1947; Dahms and Kagan 1938; Eyles 1963; Hesse 1947; Kirkpatrick 1933; Lent 1939; Miller 1957; Milliken and Wadley 1922; Odihambo 1957; Poisson 1930; Putshkova 1956; Schneider 1940; Scudder 1963; Slater 1951, 1964; Sweet 1960, 1964; Tischler 1960; Usinger 1942a; Usinger and Ashlock 1959; Woodward 1968; York 1944.

THAUMASTELLIDÆ Stys 1964, *Čas. Čs. Spol. ent.* (*Acta Soc. ent. Cechoslov.*) **61,** 3, 238–253. (Fig. 13)

In 1896 Horvath described from a single specimen an aberrant new genus and species under the name *Thaumastella aradoides* which he placed in the Lygæidæ (*Termes Fuzetek* **19,** 325). Seidenstucker (1960, *Stuttgarter Beitraege zur Naturkunde*, Stuttgart, No. 38, 1–4) identified a specimen of this species from Iran and redescribed and figured it, establishing a new subfamily of Lygæidæ for it—**Thaumastellinæ.**

In 1964, Seidenstucker transferred *Thaumastella* to the Cydnidæ. Stys (*Acta Soc. ent Czechoslovakia* **61,** 238) redescribed the species and raised Seidenstucker's subfamily of Cydnidæ to family rank Thaumastellidæ.

Members of this family are small bugs of Lygæid appearance. The principal characters are—head porrect with ocelli; antennæ with five segments; rostrum with four segments, the first segment hidden in the buccular groove; scutellum triangular, longer than short claval commissure; corium divided by medial fracture into exo- and endocorium, both parts separately rounded posteriorly; membrane with non-recognizable venation; metathoracic wing without a distinct hamus, with both cubital furrows and secondary veins without anal veins; stridulatory apparatus present; strigil

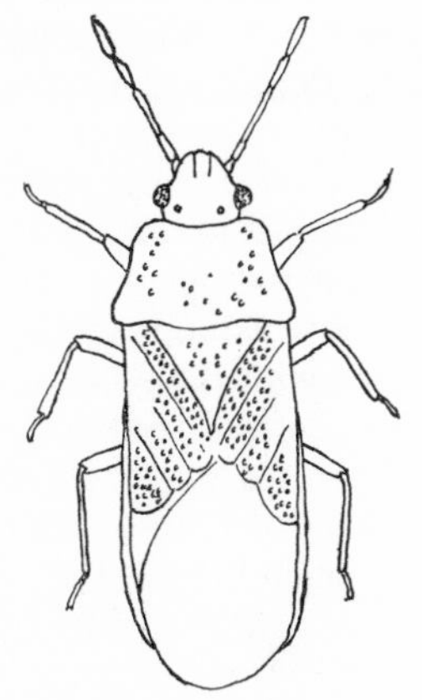

Fig. 13

Thaumastella aradoides Horvath 1896 (after Seidenstucker) (Thaumastellidæ).

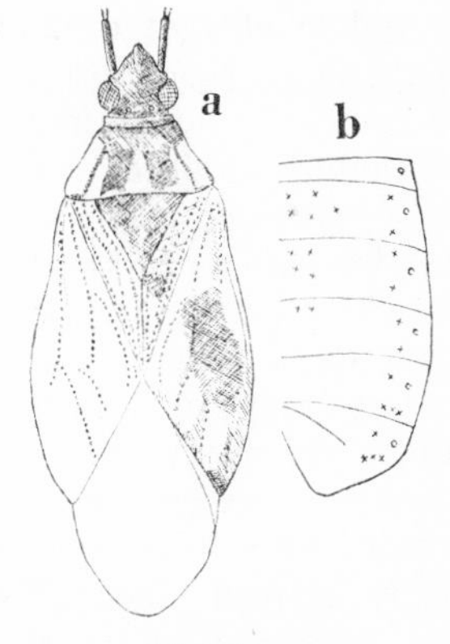

Fig. 14

Idiostolus insularis Berg 1884. A. dorsal view. B. half ventral view with position of trichobothria and spiracles (after Scudder and China) (Idiostolidæ).

situated between cubital furrow and Pcu, the plectrum formed by the paired limæ on second abdominal tergum; the third tergum with sharply delimited medial circular region; dorsal glands absent; seventh ventrite of female not split; pygophore posteriorly truncate; phallus apparently simple; eighth ventro-lateral tergites fused into a transverse plate situated between the eighth and ninth tergum;

all abdominal spiracles ventral; legs without spines; tarsi with three segments.

They represent a relict group which diverged from the pentatomine stock at an early stage of the evolution of the superfamily Pentatomoidea.

The single species *T. aradoides* Stys was originally described from Algeria. It occurs also in Iran, the Sudan and Iraq and doubtfully in Saigon.

Nothing is known, apparently, about its biology.

IDIOSTOLIDÆ Stys 1964, *Acta Soc. ent. Czechoslovakia* **61,** 238 (Fig. 14)

The subfamily **Idiostolinæ** Scudder 1962, *Canadian Entomologist* **94,** 1066–1069, was elevated to family rank by Stys 1964. Up to the present only one species has been recorded, namely *insularis* Berg (1884, *Addit. Hem. Argent.* 61) from Tierra del Fuego, Argentine. Nothing is apparently known about its ecology.

The principal characters of the family are as follows: antennæ with four segments inserted below the line joining apex of head and middle of eye; ocelli present; rostrum with four segments; bucculæ short; head more or less impunctate; thorax with pronotum, scutellum and sterna impunctate; scutellum triangular with the apex slightly produced and spatulate; pro-, meso-, and metasterna clearly separate; posterior margin of metapleura straight and somewhat laminate; pleural sulcus visible on mesopleura; odoriferous gland ostiole and peritreme distinct; legs without spines and teeth; femora slender, not at all swollen; tarsi with three tarsomeres and large fleshy arolia; hemielytra with clavus, corium and membrane distinct, but no cuneus or embolium; punctures along all longitudinal veins except the one adjacent to anterior (lateral) margin; venation with five longitudinal veins to membrane, the veins apically ramate and anastomosing; metathoracic wings with

---

Fig. 15 (*facing*)

Neanides of Plataspidæ, Pentatomidæ, Coreidæ, Pyrrhocoridæ, Piesmatidæ

1. *Chrysocoris stockerus* Linnæus 1764. Pentatomidæ-Scutellerinæ. 4th instar.
2. *Brachyplatys vahlii* (Fabricius) 1787. Plataspidæ. 1st instar.
3. *Eusthenes robustus* Lepeletier and Serville 1825. Pentatomidæ-Tessaratominæ. 1st instar.
4. *Plataspis* sp. Plataspidæ. 4th instar.
5. *Antilochus nigripes* Burmeister 1835. Pyrrhocoridæ. 2nd instar.
6. *Physomerus parvulus* Dallas 1851. Coreidæ-Coreinæ. 1st instar.
7. *Piesma quadratum* Fieber 1861. Piesmatidæ. 5th instar.
8. *Gonopsis pallescens* Distant 1902. Pentatomidæ-Phyllocephalinæ. 4th instar.
9. *Ectatops* sp. Pyrrhocoridæ. 4th instar.

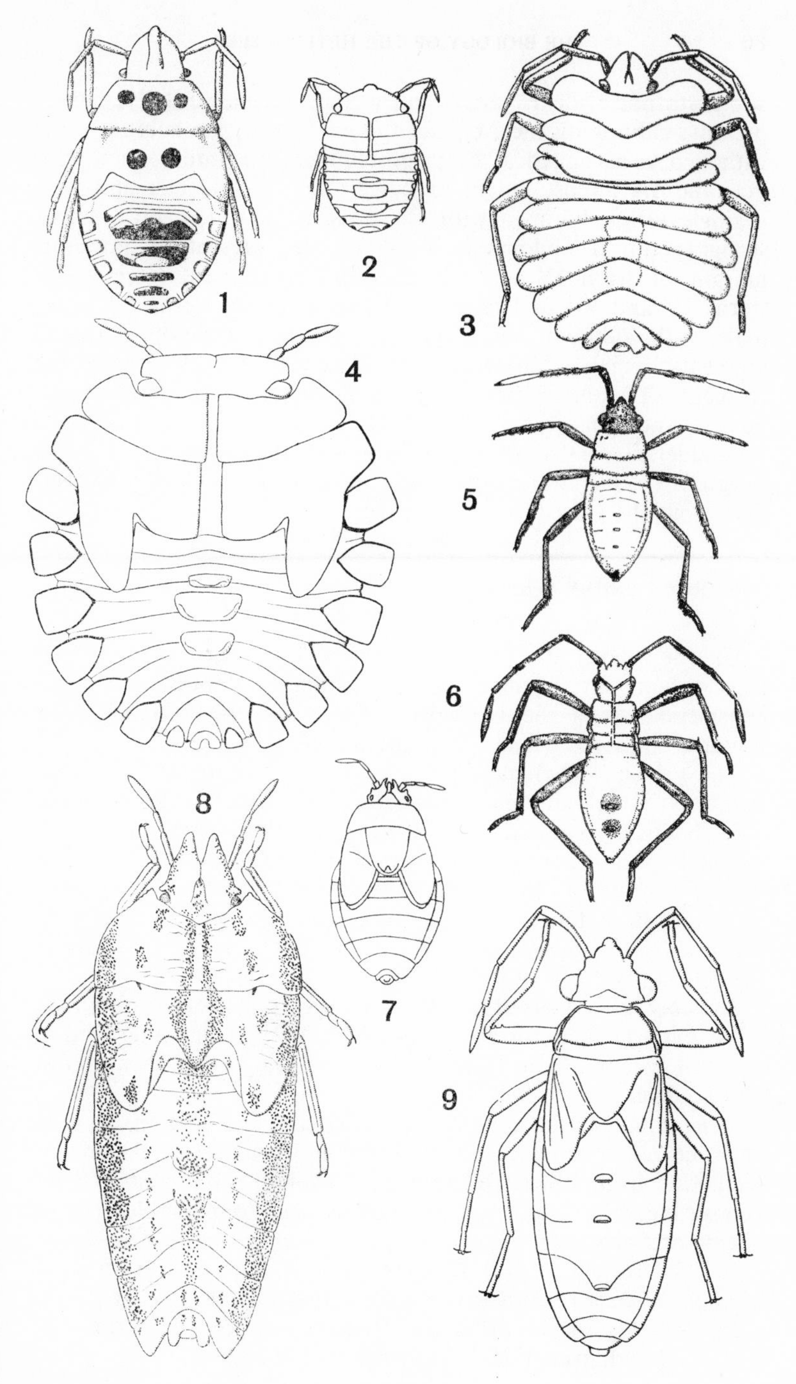

Fig. 15

sub-costa not evident; radius and median divided distally; hamus present; cubitus distinct; cubital furrow forked, the cubital sector with two veins united at base; postcubital area rather small with Pcu and IA almost united at base; anal lobe distinct and with a single vein; abdomen with all spiracles ventral; ventral sulcus straight; dorsal abdominal gland ostioles present at posterior margin of terga IV and V; trichobothria present laterally on sterna IV and VII and laterally and medially on sterna III to V; male with sternum VII large and partially covering terminal segments; female with sternum VII cleft midventrally, overlapping segment VIII and sometimes partially divided into anterior and posterior portions.

Scudder associated *Idiostolus* Berg with the Australian *Trisecus* Bergroth (*T. pictus* Bergroth 1893) to form the subfamily **Idiostolinæ,** which he placed in the Lygæidæ.

Reference Schaefer 1966.

PYRRHOCORIDÆ Amyot and Serville 1843, *Hist. nat. Hémipt.* xxviii, 265 (Plate 3)

On the whole, the Pyrrhocoridæ are larger and more robust insects, and on account of their brighter colours, are more conspicuous than the Lygæidæ. There are approximately four hundred species and their distribution is world-wide.

The principal characters are: ocelli absent; antennæ with four segments; scutellum small, triangular; hemelytra usually complete with clavus, corium and membrane; tarsi with three segments.

Some species, namely *Melamphaus faber* (Fabricius) 1787 and several species of *Dysdercus* Amyot and Serville 1843, are of economic importance. *M. faber* has been recorded as damaging the seeds of *Hydnocarpus* in Malaya, and *Dysdercus* are pests of the cotton plant and other Malvaceæ.

M. faber was observed to show a marked preference for the fruit of *H. anthelmintica*, for the probable reason that the outer rind of the fruit is only 2 mm. thick, and is much less woody than that of *H. wightiana*.

When handled, the adults of *M. faber* secrete a copious quantity of a very objectionable and colourless fluid, which, unlike similar secretions from other Heteroptera does not leave a stain after volatilizing. The odour of the fluid is said to resemble that of phenyl-acet-aldehyde.

It is interesting to note that the smell of the immature nuts of *H. anthelmintica* resembles very closely that of stale raw meat.

Incidentally, the oil from the *Hydnocarpus* seeds was formerly used (and perhaps still is) in the treatment of leprosy.

The ova of *M. faber* are regularly ovate, very feebly reticulate, smooth, shining opalescent and are deposited in loose masses.

Information concerning the ecology and developmental stages of the Pyrrhocoridæ is very scanty, since only those damaging economic crops have been studied to any extent.

It would seem that the ova are deposited in large groups in soil and leaf mould. Ova which have been examined are ovate with the surface of the chorion smooth and shining. Eclosion is effected by the embryo splitting the chorion more or less along its longer axis. The adults and neanides in subtropical regions conceal themselves in leaf-debris and under logs or bark during the cold season.

Certain species of *Dysdercus* attack plants other than cotton. For example, *Dysdercus cingulatus* (Fabricius) 1775 has been reported as attacking bottle gourd (*Lagenaria vulgaris*), musk mallow (*Hibiscus abelmoschus*) and cabbage (*Brassica oleracea*).

The Pyrrhocoridæ appear to be mainly phytophagous, but among them, at least two carnivorous species have come to notice. They are *Dindymus rubiginosus* (Fabricius) 1787, a species resembling *Dysdercus* which attacks larvæ of *Oreta extensa* Walker (Lepidoptera-Drepanidæ) and adults of *Lawana candida* (Fabricius) (Homoptera-Flatidæ) and *Antilochus coqueberti* (Fabricius) 1803, killing *Dysdercus cingulatus*.

M. faber has been observed to feed on its own ova and also to attack neanides in the laboratory.

Most of the Pyrrhocoridæ are fully alate. There are, however, some apterous genera, namely, *Courtesius* Distant 1903 from India and *Myrmoplata* Gerstaecker 1892, which occurs in the Ethiopian and Oriental Regions.

Brachyptery often occurs, e.g., in the genera *Scantius* Stål 1865 (Oriental and Ethiopian Regions), *Pyrrhocoris* Fallèn 1814 (Palæarctic Region), *Cenæus* Stål 1861 and *Dermatinus* Stål 1853 (Ethiopian Region).

References

China 1954; Kirkaldy 1900; Miller 1932a; Whitfield 1933.

LARGIDÆ Dohrn 1859, *Cat. Hem.* 36 (Plate 3)

The Largidæ are separated from the Pyrrhocoridæ by having the seventh ventral segment of the female medially divided. Their habits are similar to those of the Pyrrhocoridæ, so far as is known.

The very large and sexually dimorphic *Lohita grandis* Gray 1832, is of some economic importance. *Euryopthalmus sellatus* (Guérin) 1857 is apparently carnivorous.

Two somewhat remarkable species, both from the point of view of colouration and structure, are *Astemma stylopthalma* Stål 1870, a black species with a bright red corium and pedunculate eyes, and *Fibrena gibbicollis* Stål 1861, in which the male has the anterior lobe of the pronotum gibbose. The female is of normal shape.

Brachyptery occurs in *Arhaphe* Herrich-Schaeffer 1853, *Japetus* Distant 1883, *Phæax* Distant 1893 and *Stenomacra* Stål 1870, all Neotropical, except *Arhaphe* which is distributed in the Palæarctic Region and in the Sonoran sub-region. It should be noted that the name *Japetus* is preoccupied. (Homoptera.) *Lohita grandis* has been recorded from Burma, Cochin China and Sumatra.

So far as is known, oviposition methods are similar to those of the Pyrrhocoridæ. An ovipositor is present in *Euryopthalmus* which may indicate a different mode of oviposition. The Largidæ are probably more nearly related to the Lygæidæ than to the Pyrrhocoridæ.

Reference Lattin 1958.

PIESMATIDÆ (Amyot and Serville) 1843, *Hist. nat. Hémipt.* xl. (Fig. 16)

A family containing three genera only. It is characterized by having the scutellum exposed, ocelli, the corium, clavus and base of membrane areolate, tarsi with two segments, membrane in macropterous forms with four straight veins. Most of the species, all of which are phytophagous, are Palæarctic, and they number about twenty. There are also some species distributed in the Nearctic and Ethiopian Regions. The genus *Miespa* Drake 1948 has been recorded in Chili, and *Mcattella* Drake 1924 from Australia.

A Palæarctic species *Piesma quadratum* Fieber 1861 has been extensively studied by Wille. Its ova which are about 0·60 mm. in length, are usually deposited with the longer axis parallel to the substratum. An operculum such as found in the ovum of *Cimex lectularius* Linnæus 1758 (Cimicidæ-Cimicinæ) is present. The ovum is provided with five or six processes for the purposes of aeration.

The operculum, however, on eclosion, splits in five or six directions and is not removed entire by the emerging neanide (Fig. 9, 7), and the cuticle which had enveloped it remains attached to the chorion.

P. quadratum is an important pest of beet. *P. cinereum* (Say) 1832 has been recorded as being a vector of a virus disease affecting sugar-beet in N. America. The disease is known as "Savoy". Affected beet plants exhibit dwarfed, down-curled leaves, the most conspicuous being the innermost.

References

Coons, Cotila and Stewart 1937; Drake and Davis 1958; Leston 1954a; Schneider 1928; Wille 1929.

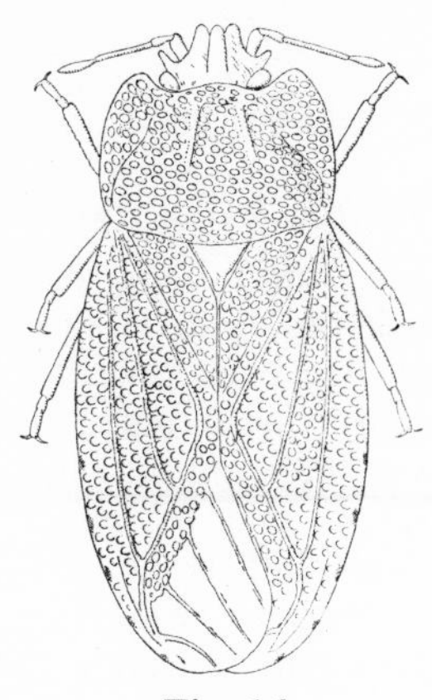

Fig. 16

Piesma dilutum Stål 1855 (Piesmatidæ).

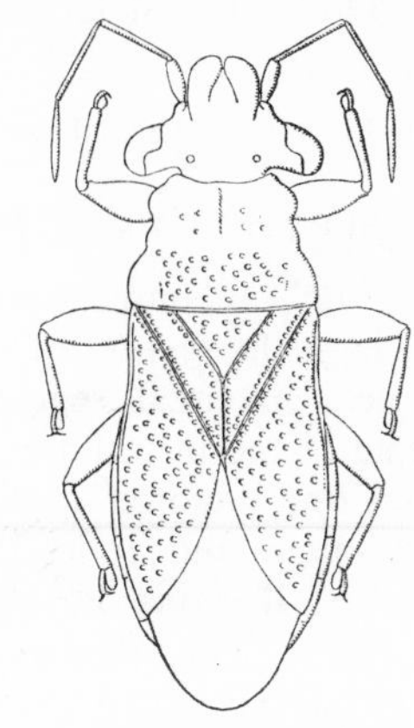

Fig. 17

Thaumastocoris australicus Kirkaldy 1908 (Thaumastocoridæ).

THAUMASTOCORIDÆ Kirkaldy 1908, *Proc. Linn. Soc. N.S.W.* **32,** 789 (Fig. 17)

This family is divided into two subfamilies, **Thaumastocorinæ** Kirkaldy 1908 and **Xylastodorinæ** Barber 1920. In the former the pseudarolia are absent, the apex of the tibiæ has a lobate, sensory appendage; in the males one harpago is present and the pygophore has a lateral projection; the apex of the corium is remote from the posterior apex of the membrane. In the latter, pseudarolia are present and the apex of the tibiæ lack a lobate sensory appendage; in the males, harpagones and the lateral projection on the pygophore are absent; the lateral margin of the corium extends nearly to the apex of the membrane.

Distribution embraces Australia, Southern India, Argentine and the Greater Antilles. It appears to be a relict group but not a primitive one. Knowledge of the natural geographic range of the genera and species is probably very incomplete.

The food of two species of Xylastodorinæ consists of the newly developing and unfolding terminal leaves of palms. *Xylastodoris luteolus* Barber 1920 feeds on the Royal Palm (*Oreodoxa regia* H. B. and K.) and *X. vianai* Kormilev 1955 on *Euterpe edulis* Mart.

('Palmito'). *Onymocoris hackeri* Drake and Slater 1957 has been collected from *Banksia* sp. in Australia.

The ova of *X. luteolus* are deposited among the membranous scales covering the underside of the leaflet midrib.

A stridulatory furrow has been detected in *Thaumastocoris hackeri* Drake and Slater 1957.

References

Baranowski 1958; Slater and Drake 1956; Drake and Slater 1957.

BERYTIDÆ Fieber 1851, *Genera Hydroc.* 9 (Fig. 18)

Small, delicate insects with an elongate linear body and long slender legs, the femora of which are thickened apically. The basal and apical antennal segments are also thickened apically. The peritreme of the metathoracic gland ostioles is produced.

About one hundred species are known; these are distributed in the Ethiopian and Indo-Australian Regions and belong to two subfamilies, **Berytinæ** Puton 1886 and **Metacanthinæ** Douglas and Scott 1865. The **Berytinæ** have an elongate head with the vertex produced above the tylus; the **Metacanthinæ** have a shorter head and no process on the vertex.

The Berytidæ are found mainly among grasses and on tree-trunks. *Gampsocoris pulchellus* (Dallas) 1852 **(Metacanthinæ)** in both the neanide and adult stages has been recorded on cacao, the younger shoots of which they attack. The female deposits its ova singly among the shoots. The ovum is whitish, cylindrical, somewhat curved rounded at each end and with the surface striate.

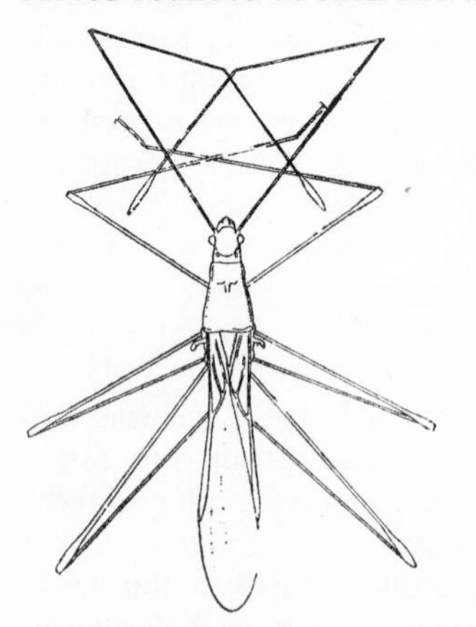

Fig. 18
Metacanthus pertenerum Breddin 1907 (Berytidæ).

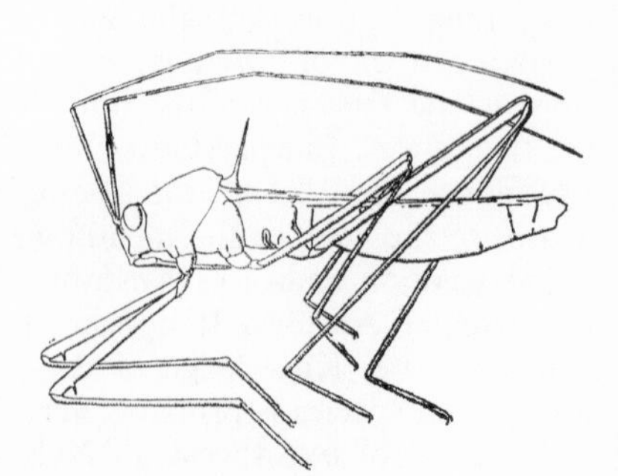

Fig. 19
Phænacantha suturalis Horvath 1904 (Colobathristidæ).

This species has also been found on *Hibiscus mutabilis* and on *Passiflora fœtida*. Adults have been seen to be attacked by adults of *Cosmolestes picticeps* Stål 1859 (Reduviidæ-Harpactorinæ).

The ovum of *Metatropis rufescens* Herrich-Schaeffer 1835 **(Metacanthinæ)** is cylindrical tapering posteriorly and with four minute button-shaped processes or tubercles at the anterior end. It may be attached to the substratum by a small, stout support or stalk, but not in all cases.

On eclosion, the neanide causes the chorion to split longitudinally at the anterior end. In relation to the size of the adult the size of the ovum is very small.

References

Massee 1949; Miller 1941.

COLOBATHRISTIDÆ (Stål) 1866, *Hem. Afr.* **2,** 121 (Fig. 19)

Slender, elongate Heteroptera with the head almost vertical, long and slender antennæ, a sculptured thorax and scutellum bearing spines or tubercles, long legs, large eyes and ocelli. Brachypterous forms occur in the genus *Trichocentrus* Horvath 1904.

About seventy species are known. These are distributed in the Neotropical and Indo-Australian Regions. Very little is known about their ecology.

The ovum of a species of *Phaenacantha* Horvath 1904 from Borneo is cylindrical with ten or more short chorionic processes and some short tubercles on the chorion. This information is based on the study of fragments of a dissected ovum.

Phænacantha saccaricida (Karsch) 1888 is a pest of sugar-cane in Java. The adults and neanides feed on the undersides of the cane leaves. Ova are deposited mostly in damp places singly and close to, or in the soil.

References

van Deventer 1906; Kalshoven 1950; Stys 1966.

ARADIDÆ (Spinola) 1837, *Essai Hemipt.* 157 (Figs. 20, 21)

Members of the family Aradidæ are all of dull colouration. Some of them are brachypterous and there are many apterous species. Although the alate forms are relatively active and sometimes, during flight, are attracted to artificial light, in the main they are photophobic and pass most of their existence in concealment, often under the loose bark of decaying trees. They are also found on foliage.

Their food appears to be mainly fungi, the mycelia of which are usually abundant in decaying timber. To be able to follow the ramifications of the mycelia, the stylets of the Aradidæ are admirably suited, being very long, in fact, in some species, longer than the head and body together. The rostrum, on the other hand, is usually relatively short. When the stylets are not in use, they are coiled in the anterior part of the head.

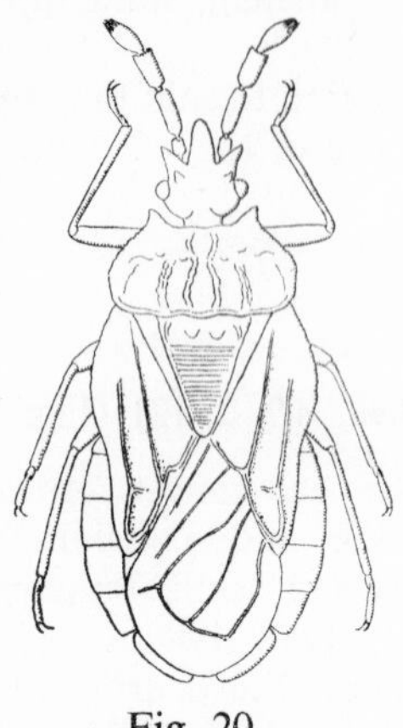

Fig. 20
Aradus depressus (Fabricius) 1803
(Aradidæ-Aradinæ).

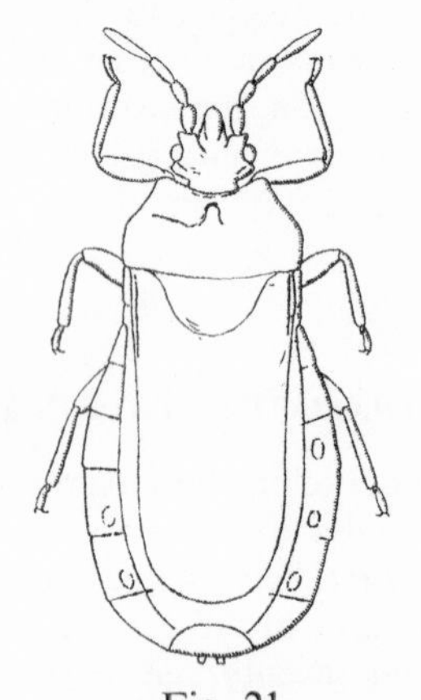

Fig. 21
Aneurus lævis (Fabricius) 1803
(Aradidæ-Aradinæ).

An interesting feature of some of the apterous genera is the thick tomentose clothing or a tenuous covering of a resin-like substance, both of which conceal to a marked degree the actual habitus and also the somewhat complicated sculpturation which adorns the dorsal surface of the thorax and abdomen.

Not a great deal is known about the ecology of the Aradidæ. Ova are usually deposited under bark. The ovum of *Aradus depressus* (Fabricius) 1803 is oval, white with a faint greenish tinge; there is a ring of low elevations at the anterior end. The embryo ruptures the chorion within this ring. The ova are attached by the side and are deposited singly.

Scent glands are present in most species.

There are eight subfamilies. **Isoderminæ** Stål 1872, in which the rostrum is free at base and the wings deciduous at a line of weakness level with the apex of the scutellum. This is a small subfamily of strongly flattened insects distributed in Chile, New Zealand, Australia and Tasmania. **Prosympiestinæ** Usinger and Matsuda 1959, also a small subfamily of primitive Aradids in which the odoriferous gland ostioles have a curved seta and the well-developed juga extend on each side of the clypeus for nearly half its length.

Distributed in Australia, Tasmania and New Zealand. **Chinamyersiinæ** Usinger and Matsuda 1959. In representatives of this small subfamily the odoriferous gland ostioles have a carina curved at its apex and the juga reduced. Distribution, New Zealand. **Calisiinæ (Stål)** 1873. In this the scutellum is large, covering the abdomen almost entirely, the margin of the connexivum is double and the hemelytra are largely hidden and membranous. Distribution mainly in the Southern Hemisphere. **Aradinæ** (Amyot and Serville) 1843. A medium-sized subfamily in which the scutellum is small and covers only a small part of the abdomen; the hemelytra are completely exposed and often have the costal margin dilated basally. Distribution mainly Holarctic. **Aneurinæ** (Douglas and Scott) 1865. In this small subfamily, representatives have the rostrum arising from an open atrium and the anterior dorsal abdominal odoriferous gland ostiole of the third segment not or slightly displaced backward. Distribution, cosmopolitan. **Carventinæ** Usinger 1950. Body more or less covered with a pale incrustation which forms a constant pattern and sometimes entirely obscures the integument. Distribution, Ethiopian, Australian, Oriental, Neotropical and Nearctic Regions. **Mezirinæ** Oshanin 1908. In this subfamily the rostrum does not extend beyond the base of the head, the prosternum is not sulcate, the head is wider behind the eyes than in front and the spiracles are equidistant between the basal and apical margins of segments. The metathoracic odoriferous gland ostioles are distinct, placed laterally behind the median coxæ and usually narrow-elongate with a channel leading from the inner end towards the median coxæ. Distribution, world-wide.

References

Barber 1933; Bueno, de la Torre 1935; Butler 1923; Jordan 1932; Kiritshenko 1913; Miller 1938; Tamanini 1956; Usinger 1941, 1950, 1959; Wygodzinsky 1946b.

TERMITAPHIDIDÆ Myers, *Psyche Camb. Mass.* **31,** 6, 267 (Fig. 22)

The Termitaphididæ are very small insects, elliptical in outline, without eyes or ocelli and with the lateral margins of the thorax laminate, the external margin of each lamella having tubercles and flagella.

Apart from a few details regarding the ova of *Termitaradus trinidadensis* (Morrison) 1923, little is known about the ecology of these small and obscurely coloured insects.

In 1902 a curious termitophilous insect was discovered and, at the time, was considered to be an abnormal aphid. For this insect, Wasmann erected the genus *Termitaphis* but, nine years later, Silvestri, when describing two more new species, came to the conclusion that the genus did not belong to the Homoptera. He therefore placed it in the Heteroptera and established a new family, the Termitocoridæ to receive it. On account of the fact, however, that a family name must be derived from the type genus (the genus *Termitocoris* being non-existent) the name of the family must be Termitaphididæ.

The species known up to the present, ten in all, have been found in the nests of termites, but it has not been definitely established what their food is.

An examination of the stylets has revealed a resemblance to those of the Aradidæ for they are very long and spirally coiled in the head capsule. It is assumed, as in the Aradidæ, that these elongate mouth-parts are adapted for feeding on the mycelia of fungi. On the other hand, it is not improbable that the Termitaphididæ may be carnivorous and with their long stylets may be able to reach such insects as coleopterous larvæ feeding below the surface.

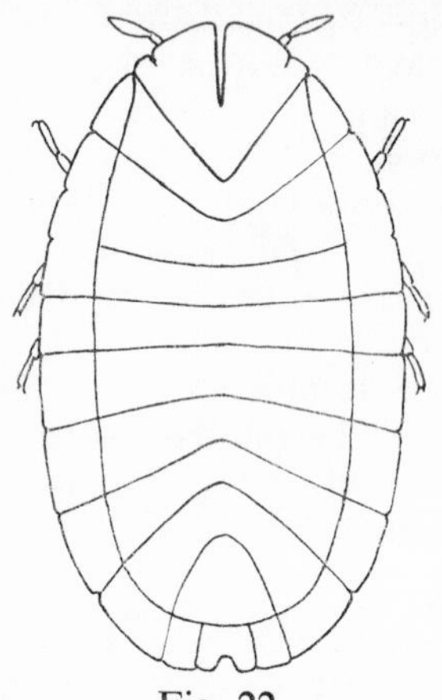

Fig. 22
Termitaradus panamensis Myers 1924 (Termitaphididæ).

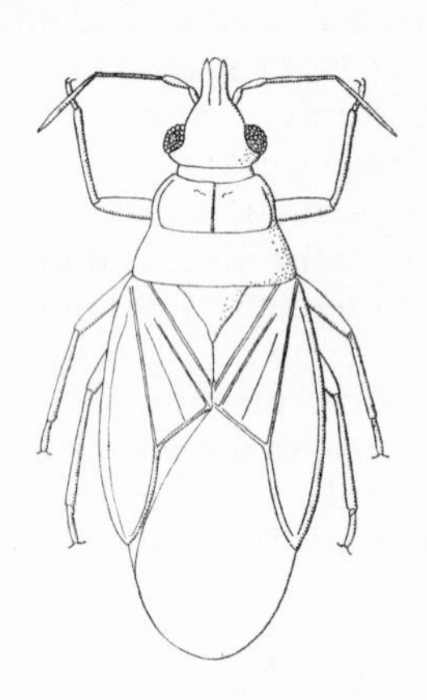

Fig. 23
Joppeicus paradoxus Puton 1898 (Joppeicidæ).

The ova of *Termitaradus guianæ* (Morrison) 1923, are described as follows: The surface is smooth, extremely polished and porcelain-like. Under very strong direct light it is possible to discern a faint and somewhat irregular pitting. There is not the slightest external sign of micropylar apparatus or cap; nor does there appear any difference in the evanescent pitting in different parts of the chorion. The dimensions of the ova which are ovoid are 0·86 mm. ×0·56 mm.

Termitaphididæ have been recorded from Central America, the West Indies, India and Africa.

References

Mjöberg 1914; Morrison 1923; Myers 1924, 1932; Silvestri 1911, 1921; Usinger 1942b.

JOPPEICIDÆ Reuter 1910 *Acta Soc. Sci. Fenn.* **37,** 3, 75 (Fig. 23)

Small insects with the rostrum, which is composed of four segments, directed forwards, short legs and tarsi with two segments. The hemelytra have a strongly sclerotized corium and larger membrane with four nervures. They are related, according to China, to the Reduviidæ.

There is one genus, *Joppeicus* Puton 1881, found in Syria, but very little is known about it. It occurs on *Ficus sycomorus* where, no doubt, it lives as a predator on other small insects.

Neanidal *Joppeicus* have three abdominal glands. The segmentation of the abdomen being abnormal, however, the ostioles are on the fourth, fifth and seventh segments towards the anterior margin but separated from it.

Reference

China 1955a.

TINGIDÆ (Costa, A.) 1838 *Cimicum Regni Neap. Cent.* **1,** 18 (Figs. 24, 46)

Mostly very small insects including several genera with striking modifications to the pronotum and hemelytra, the former usually concealing the scutellum; hemelytra are densely reticulate or areolate.

Several species are of economic importance, occasionally causing defoliation or causing galls to form, for example, certain species of *Copium* Thunberg 1822, —*cornutum* Thunberg 1822, *teucrii* (Host) 1788, *brevicorne* (Jakovlev) 1879, *magnicorne* (Rey) 1888, *reyi* Wagner 1954 and *japonicum* Esaki 1931, are associated with Labiatæ into the flowers of which they oviposit and it is thought that a substance produced during oviposition or arising from the ova is responsible for abnormal proliferation of the plant tissue.

A probable case of facultative blood-sucking has been recorded for *Corythuca cydoniæ* (Fitch) 1861.

On the whole not a great deal is known about the ecology of the Tingidæ. According to observations, ova have been found on leaves, usually the underside and sometimes they are inserted into the tissue or are covered with a sticky substance. Myrmecophilism occurs in the Tingidæ, for example, adults and neanides of *Allocader leai* (Hacker) 1928, **Tinginæ,** have been found in the nest of the ant *Amblyopone australis* Erich in Tasmania and adults of *Lasiacantha leai* (Hacker) 1928, **Tinginæ,** in the nest of *Iridomyrmex conifer* Forel in Western Australia. Both of these inquilines have fully-developed compound eyes.

The family is divided into three subfamilies. **Cantacaderinæ** (Stål)

1873, in which the head is strongly produced in front of the site of insertion of the antennæ, the bucculæ are anteriorly produced and do not extend posteriorly beyond the anterior margin of the pronotum, the pronotum is posteriorly obtusely rounded or angulate, rarely truncate and disclosing the apex of the scutellum; the hemelytra have the discal cell usually divided by a transverse vein or veins into 2 or 3 areas. Distribution, Old World. **Agramminæ** (Douglas and Scott) 1865. In this subfamily the hemelytral areolations are very small, the corium lacks distinct discoidal and sub-costal areas, the costal margin of the hemelytra is not dilated and the pronotum has neither a vesiculate hood nor lateral paranotal dilations. Distribution, world-wide. **Tinginæ** (Douglas and Scott)

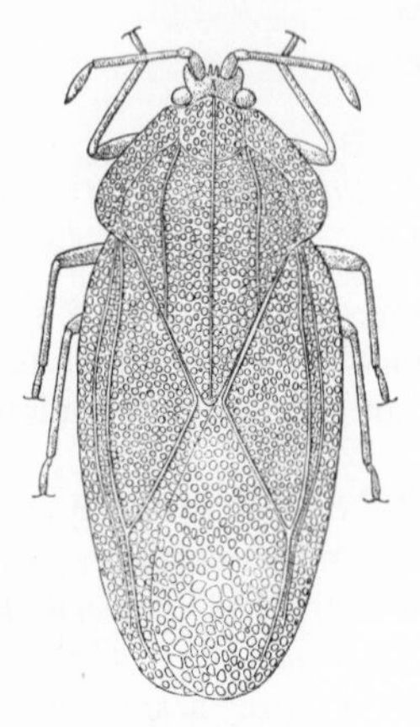

Fig. 24
Tingis cardui Linnæus 1785 (Tingidæ).

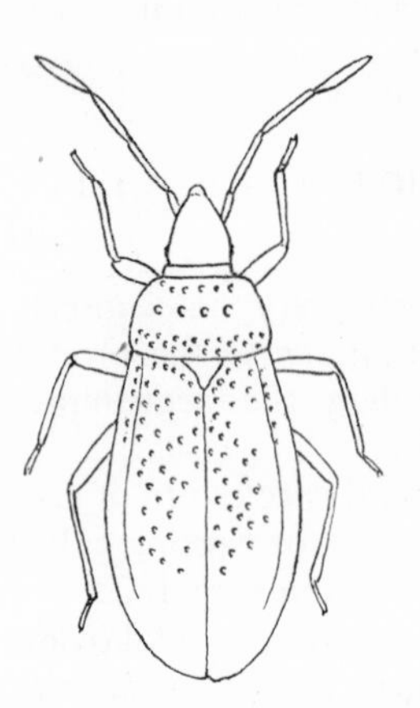

Fig. 25
Anommatocoris coleoptrata (Kormil ev) 1954-5 (Vianaididæ).

1865. This subfamily is the largest; representatives have the hemelytra with usually large areolations, the corium with distinct discoidal and sub-costal areas, the pronotum usually with a vesiculate hood or hood-like prominence, and often with paranotal dilations. Distribution, cosmopolitan.

References

Bailey 1951; Behr 1952; Buchanan White 1877; Butler 1923; Carayon 1958; Douglas 1877; Drake 1925; Drake and Davis 1960; Esaki 1931; Frauenfeld 1853; Horvàth 1906; Houard 1906; Johnson 1936; Leston 1954b; Livingstone 1962; Malhota 1958; Monod and Carayon 1958; Roonwal 1952; Rubsaamen 1895; Sailer 1945; Southwood and Scudder 1956; Thontadarya and Basavanna 1959; Stusak 1957, 1962; van der Vecht 1953.

VIANAIDIDÆ Kormilev 1955, *Rev. Ecuat. Ent.* **2,** 465-477 (Fig. 25)

This family was erected to receive a hemipteron found in soil removed from their nest by leaf-cutting ants, *Acromyrmex lundi* (Guérin) and accumulated near it.

It was described as *Vianaida coleoptrata* and has the following characters-body elongate-ovate, convex on dorsal surface; head narrowed anteriorly and somewhat produced, without spines or projections; antennæ robust, long and with 4 segments; eyes rudimentary and probably not functional, consisting of six or seven facets; ocelli absent; pronotum small, wider than long, somewhat flat; evaporative area of metathoracic glands covering the greater part of the metapleura; hemelytra coriaceous, punctate, without clavus or membrane; metathoracic wings absent; spiracles small, situated on ventral surface; trichobothria absent; legs formed for running; tarsi with 2 segments and without arolia or pseudarolia. Distribution, Argentina.

Another genus and species, completely blind, was described by China as *Anommatocoris minutissimus* in 1945 and was placed in the Lygæidæ in error, as it was found subsequently. This hemipteron was found in Trinidad.

It has recently been decided that this genus and *Vianaida* are inseparable from each other, consequently they have been synonymized, thus *Vianaida coleoptrata* is now *Anommatocoris coleoptrata* (Kormilev) 1955.

A new genus and species has recently been described. This is *Thaumamannia manni* Drake and Davis 1960, and was found in the nest of an ant in Bolivia.

References

China 1945; Drake and Davis 1960; Kormilev 1955.

ENICOCEPHALIDÆ (Stål) 1960, *Rio Jan. Hemipt. I:K. Svensk. Vet-Ak. Handl.* **2,** No. 7, 1858 (Fig. 26)

The Enicocephalidæ are characterized by the elongate head which is globose behind the eyes, by the somewhat elongate habitus and the entirely membranous hemelytra. Ocelli are present except in apterous genera. The structure of the rostrum is primitive and in the genus *Ænictopechys* Breddin 1905 it is extended forwards and not folded beneath the head. Stridulatory furrow is absent. Odoriferous glands present.

Although the Enicocephalidæ are predaceous they are not provided with anterior legs of the true raptorial type, the anterior tibiæ being somewhat thick, and the tarsi are composed of one segment bearing one or two claws (mostly two) which are long in some genera.

This is one of the smaller families of the Heteroptera and contains mostly small, obscurely coloured insects, the habits of which are becoming more known than hitherto. They frequent mainly damp situations, leaf debris and accumulations of soil at the base of epiphytes. Some species have been observed to fly in the sunshine after the manner of certain Diptera.

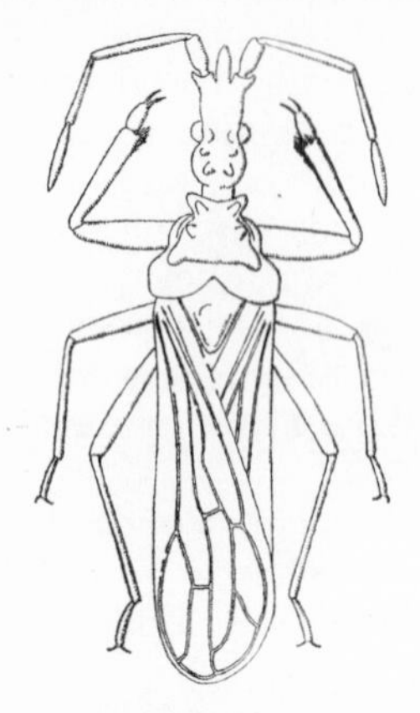

Fig. 26
Embolorrhinus cornifrons Bergroth 1905 (Enicocephalidæ).

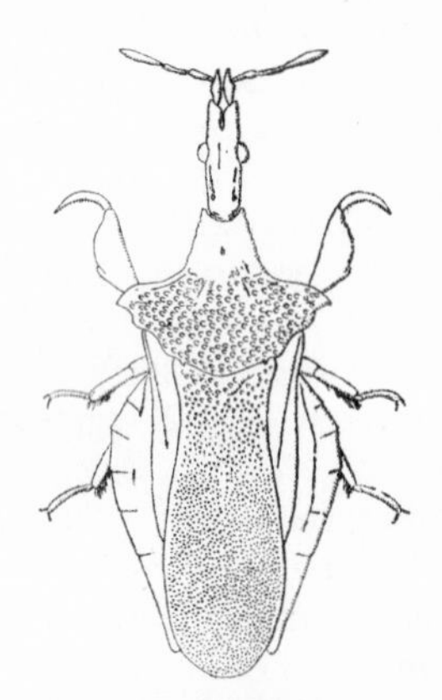

Fig. 27
Glossopelta montandoni Handlirsch 1897 (Phymatidæ-Macrocephalinæ).

With regard to the predaceous habits of these insects, it is probable that, within certain limits, they are general feeders but there is, however, a record which reveals that at least one species has a restricted diet, being myrmecophagous. This species is *Henicocephalus braunsi* Bergroth 1903, a South African species found in the nest of the ant *Rhoptromyrmex transversinodes* Mayr in Cape Colony. The opinion expressed by Bergroth that this Enococephalid is myrmecophagous was based on the fact of its considerably larger size in relation to the ant.

Hardly any information exists regarding the developmental stages. The ovum of a species of *Enicocephalus* Westwood 1837 (*Henicocephalus*) from New Zealand has been described by Myers. This ovum is elliptical with parallel sides and rounded ends. In this instance, it was affixed by the female by one side to a rootlet.

The Enicocephalidæ are cosmopolitan and are found on all the principal continents and major groups of islands.

There are two subfamilies: **Enicocephalinæ** (Stål) 1860, with the pronotum divided into 3 lobes by two distinct, transverse furrows; male pygophor with neither mobile gonopods nor a distinct anal tube. **Aenictopechinæ** Usinger 1932 with the pronotum roundly narrowing from base to apex, not divided by two distinct transverse furrows into 3 lobes, male pygophor with a pair of mobile gonopods and with a distinct anal tube. Incidentally, this last subfamily was suppressed by Usinger (1943), but has been reinstated by Villiers (1958).

References

Bergroth 1903; Carayon 1950, 1951; Jeannel 1941; Myers 1926; Usinger 1932 1943; Villiers 1958, 1962.

PHYMATIDÆ (Laporte) 1832, *Essai Classif. syst. Hémipt.* 14 (Fig. 27)

Most of the Phymatidæ are small, frequently attractively coloured insects, most of them possessing raptorial anterior legs and having the pronotum and abdomen in some species laterally expanded.

All species are predators, but, in the prolonged absence of animal food on account of unfavourable climatic conditions, they possibly have recourse to nectar or sap until their usual food is once more available.

The habits of Phymatidæ, all of which are diurnal, are the same as those of most of the Reduviidæ, and they also frequent flowers to which other insects may be attracted and consequently captured.

In the choice of a flower on which to take up its position, a Phymatid (it would appear) is not influenced by its colour but by its attractiveness to other insects. It is clear, therefore, that it does not necessarily choose for the purposes of concealment from its victim, a flower the colour of which harmonizes as closely as possible with its own colouration.

Unfortunately, there is little information regarding the ecology of the Phymatidæ with the exception of that concerning *Phymata pennsylvanica americana* Melin 1931, **Phymatinæ** (Dohrn) 1859, the life-history and habits of which have been extensively studied by Balduf.

This investigator, in discussing the prey of Phymatidæ arrived at the conclusion that the range in size of the Arthropods captured by them is very great, those recorded by him including the small and delicate Mycetophilidæ, Anthocoridæ and Miridæ and large, robust Lepidoptera of the families Pieridæ and Noctuidæ

Balduf states that the ova of *P. pennsylvanica americana* are deposited in masses of irregular shape containing varying numbers. Each mass is more or less deeply embedded in a layer of frothy substance.

The ovum of *Macrocephalus notatus* Westwood 1841, according to Wygodzinsky, is ovate with the surface which is in contact with the substratum, smooth. The female deposits them singly or in small groups without applying a glutinous substance to the exposed surface.

Readio has described and figured the ova of *Phymata erosa* Linnæus 1758 sub-species *fasciata* (Gray) 1832 (Fig. 28, 1). They, like those of *pennsylvanica*, are deposited in masses embedded individually in a frothy substance which leaves only the operculum and a portion of the chorion exposed. The female attached the mass to plant stems. The ovum is ovate with the opercular end oblique.

Phymatid ova exhibit some similarity to certain Reduviid ova in having chorial processes on the inner rim of the chorion. The methods of eclosion are similar, but, so far as is known, it has not yet been revealed whether, in the removal of the operculum, an egg-burster is present to assist the embryo.

The Phymatidæ are divided into four subfamilies: **Themonocorinæ** Carayon, Usinger and Wygodzinsky 1958, in which the members have the anterior tarsi with two segments and articulating normally with the tibia; anterior legs non-raptorial. Distribution, Ethiopian Region (Congo). **Phymatinæ** (Dohrn) 1859, in which the scutellum is short, not longer than broad and the hemelytral membrane has numerous veins extending from basal cells; the anterior tarsus is

---

Fig. 28 (*facing*) Ova of Phymatidæ, Reduviidæ

1. *Phymata erosa* sub sp. *fasciata* Gray 1832. Phymatidæ-Phymatinæ (after Readio).
2. *Stenolæmus marshalli* Distant 1903. 1·00 mm. Reduviidæ-Emesinæ.
3. *Polytoxus* sp. Reduviidæ-Saicinæ.
4. *Ptilocnemus lemur* Westwood 1840. 1·60 mm. Reduviidæ-Holoptilinæ.
5. *Stenopoda* sp. 2·80 mm. Reduviidæ-Stenopodinæ.
6. *Canthesancus gulo* Stål 1863. 2·20 mm. Reduviidæ-Stenopodinæ.
7. *Pygolampis* sp. 1·30 mm. Reduviidæ-Stenopodinæ.
8. *Oncocephalus* sp. 1·20 mm. Reduviidæ-Stenopodinæ.
9. *Elaphocranum* sp. 1·30 mm. Reduviidæ-Salyavatinæ.
10. *Lisarda* sp. 80 mm. Reduviidæ-Salyavatinæ.
11. *Salyavata variegata* Amyot and Serville 1843. 2·10 mm. Reduviidæ-Salyavatinæ (after Wygodzinsky).
12. *Petalochirus brachialis* Stål 1858. 1·00 mm. Reduviidæ-Salyavatinæ.
13. *Alvilla* sp. 0·80 mm. Reduviidæ-Salyavatinæ.
14. *Petalochirus obesus* Miller 1940. 1·00 mm. Reduviidæ-Salyavatinæ.
15. *Inara alboguttata* Stål 1863. 1·90 mm. Reduviidæ-Reduviinæ.
16. *Centraspis imperialis* Westwood 1845. 3·50 mm. Reduviidæ-Ectrichodiinæ.
17. *Platymerus* sp. 40.00 mm. Reduviidæ-Reduviinæ.
18. *Psyttala horrida* Stål 1865. 3·50 mm. Reduviidæ-Reduviinæ.
19. *Inara iracunda* Miller 1940. 2·00 mm. Reduviidæ-Reduviinæ.
20. *Cerilocus dohrni* Stål 1858. 2·60 mm. Reduviidæ-Reduviinæ.
21. *Pasiropsis vidua* Miller 1954. 1·80 mm. Reduviidæ-Reduviinæ.
22. *Archilestidium cinnabarinum* China 1925. 1·20 mm. Reduviidæ-Reduviinæ.
23. *Carcinomma simile* Horvath 1914. 1·00 mm. Reduviidæ-Cetherinæ.
24. *Sphedanocoris sabulosus* Stål 1833. 1·20 mm. Reduviidæ-Reduviinæ.
25. *Cheronea* sp. 1·30 mm. Reduviidæ-Reduviinæ.

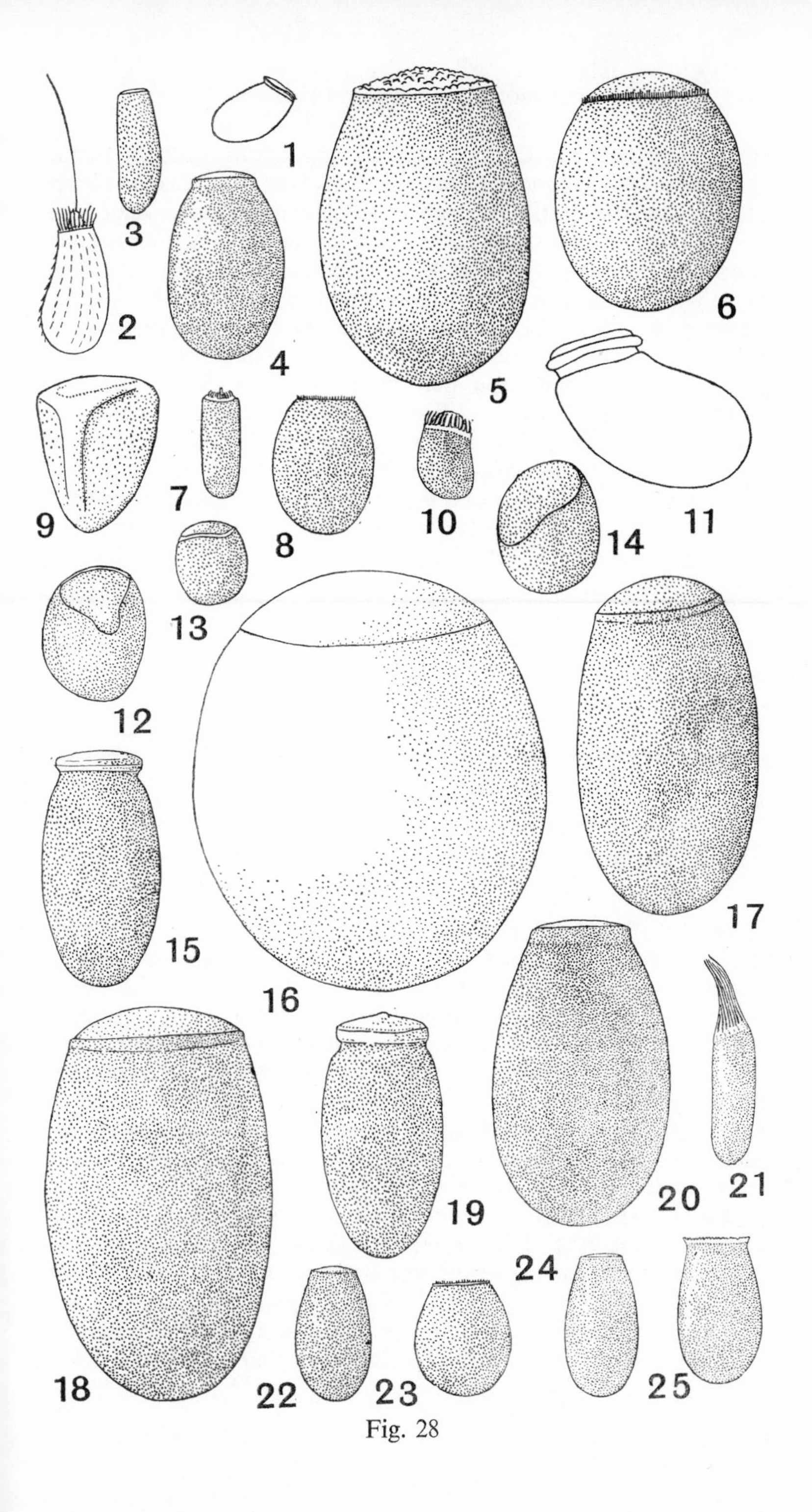

Fig. 28

fused with the tibia, the two together articulating with the femur to form a raptorial organ. Distribution, Holotropical. **Carcinocorinæ** Handlirsch 1897; in this subfamily the scutellum is longer than wide at the base, often extending to the apex of the abdomen, rarely short and only slightly longer than wide at base, in which case the hemelytral membrane has not more than five longitudinal veins; the anterior legs have the apex of the femora on the posterior side produced, so that the extended portion and the tibia act as a pincer; head and pronotum spinose. Distribution, Oriental Region. **Macrocephalinæ** (Amyot and Serville) 1843, in which the scutellum is of a similar type to that of the **Carcinocorinæ,** but the anterior legs are like those of the **Phymatinæ;** the anterior tibiæ have a row of pegs on the inner surface and the femora two parallel rows or one irregular row; when these two parts of the leg are drawn together the pegs on the tibiæ fit between the rows on the femora. Distribution tropical America.

A stridulatory furrow is present in *Glossopelta* Handlirsch 1897, *Agreuocoris* Handlirsch 1897, *Amblythyreus* Westwood 1841, *Macrocephalus* Swederus 1787 and *Narina* Distant 1906 **(Macrocephalinæ),** *Carcinocoris* Handlirsch 1897, *Carcinochelis* Fieber 1861 **(Carcinocorinæ)** and *Phymata* Latreille 1802 **(Phymatinæ).**

References

Balduf 1939, 1941; Cook 1897; Readio 1927; Wygodzinsky 1944.

ELASMODEMIDÆ Letheirry and Severin 1896, *Cat. Hémipt.* **3,** 49 (Fig. 29)

Very small and strongly dorso-ventrally compressed insects with the head transverse, divided by an arcuate transverse impression behind the eyes. The ocelli are distinct and located near the lateral margin of the postocular and adjacent to the eyes. The rostrum is very short, broad and curved. The pronotum is transversely impressed near the middle and the pro- and metathoracic acetabula project and are visible from above. The scutellum is short, wide and unarmed. The sterna are entirely flattened with a wide continuous plate-like surface and the coxæ are very widely separated. A short stridulatory furrow is present. The anterior legs are not of the raptorial type and the tibiæ are simple. The hemelytra have simple venation with RM unbranched in the corium, Cu simple and unbranched and the membrane with three simple veins which do not form cells and do not reach the apical margin. The vestiges of the neanidal abdominal gland are to be seen on the fourth segment.

Elasmodema erichsoni Stål 1860 has been found in Brazil living under loose bark of fencing posts. *E. setigerum* (Usinger) 1943 has been recorded from Paraguay and Brazil and was discovered in a bird's nest along with other Arthropods. In the habitat of *E. erichsoni* a number of pseudoscorpions, Hemiptera of the families Aradidæ and Anthocoridæ, also Dermaptera, Coleoptera, dipterous larvæ, but chiefly Psocidæ, were present.

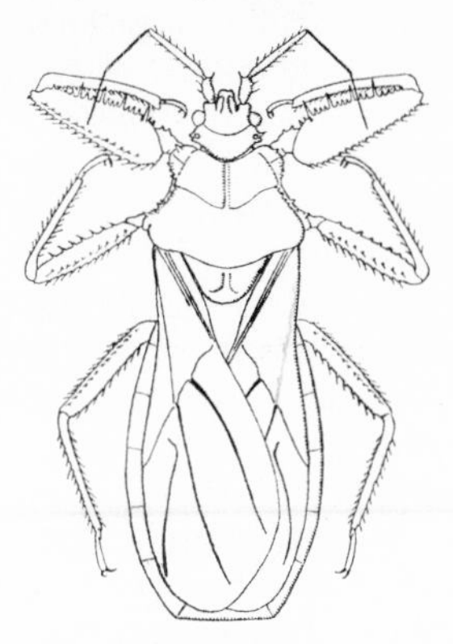

Fig. 29
Elasmodema erichsoni Stål 1860 (Elasmodemidæ).

Another species, *E. bosqui* Kormilev 1948, has been found in the Argentine. *Elasmodema* species have strong thigmotactic tendencies but they are not markedly photophobic.

The ova of *E. erichsoni* are deposited by the female in irregular groups or singly and are always fixed to the timber and not to the internal face of the bark enclosing it.

The ovum is cylindrical. The chorion is finely sculptured with pentagons and hexagons. The number of ova deposited by a single female is unknown, but, in view of its relatively large size—1·50 mm. —and the size of the female, it is likely that very few ova develop at one time.

References

Kormilev 1948; Usinger 1943; Wygodzinsky 1944.

REDUVIIDÆ Latreille 1807, *Gen. Crust. Ins.* **3,** 126
(Plates III and IV) (Figs. 28, 30, 33, 45–47)

The Reduviidæ form one of the largest families of the Heteroptera. So far as is known, all are exclusively predaceous. Representatives of the **Triatominæ** feed on mammalian and avian blood. Some of them are vectors of human trypanosomiasis. The family has the following characters: body more or less elongate, sometimes bacilliorm or linear: head usually with a transverse sulcus behind or between the eyes; ocelli present, except sometimes in apterous forms; antennæ with four to eight segments; exceptionally with as many as forty segments; no genus with five segments, intercalary segments sometimes present in antennæ composed of four segments; stridulatory furrow generally present; rostrum nearly always composed of four (three visible) segments; anterior and median tibiæ often with a fossula spongiosa; alary polymorphism frequent.

Apart from the hæmatophagous species, Reduviidæ are not of importance economically. Being mostly general feeders they are not effective in causing an appreciable reduction in the numbers of insects which are pests of crops.

The Reduviidæ are to be found in many kinds of terrestrial habitat and are most abundant in tropical and sub-tropical regions. Some are found occasionally in caves, but there are no true cavernicolous Reduviidæ in the strict sense.

As to their habits they may, with reasonable accuracy, be divided into two categories, namely, diurnal and nocturnal. These two categories, however, are not always sharply defined. Diurnal species are found mainly on bushes, low herbage and sometimes fairly high up on the foliage. Nocturnal species remain in seclusion in the daytime, but may occasionally be seen in the open. Their appearance may be voluntary, for example, during the search for

---

Fig. 30 Ova of Reduviidæ (*facing*)

1. *Neocentrocnemis signoreti* (Stål) 1863. 3·00 mm.
2. *Psophis consanguinea* Distant 1903. 2·50 mm.
3. *Dyakocoris vulnerans* (Stål). 2·20 mm.
4. *Zelurus luteoguttatus* (Stål) 1854. 2·00 mm.
5. *Sminthus* sp. 2·50 mm.
6. *Drescherocoris horridus* Miller 1954. 1·20 mm.
7. *Holotrichius tenebrosus* Burmeister 1835. 1·50 mm.
8. *Velitra alboplagiata* (Stål). 2·10 mm.
9. *Zelurus limbatus* (Lepeletier and Serville) 1825. 1·50 mm.
10. *Durganda rubra* Amyot and Serville 1843. 1·30 mm.
11. *Acanthaspis* sp. 1·80 mm.
12. *Phyja tricolor* Distant 1919. 2·30 mm.
13. *Cethera musiva* (Germar) 1837. 0·90 mm.
14. *Acanthaspis fulvipes* Dallas 1850. 2·00 mm.
15. *Physoderes patagiata* Miller 1941. 1·00 mm.
16. *Kopsteinia variegata* Miller 1954. 1·00 mm.
17. *Velitra rubropicta* Amyot and Serville 1843. 2·00 mm.
18. *Neostachyogenys tristis* Miller 1953. 1·40 mm.
19. *Acanthaspis flavovaria* Hahn 1834. 1·50 mm.
20. *Eupheno* sp. Eupheninæ. 1·40 mm.
21. *Tiarodes nigrirostris* Stål 1859. 2·60 mm.
22. *Pteromalestes nyassæ* (Distant) 1877. 3·20 mm. Piratinæ.
23. *Ectomocoris fenestratus* Horvath 1911. 3·10 mm. Piratinæ.
24. *Rasahus sulcicollis* Serville 1831. 2·80 mm. Piratinæ.

(*Note:* Nos. 1-19 and 21 all Reduviinæ, except No. 13 Cetherinæ and No. 15 Physoderinæ.)

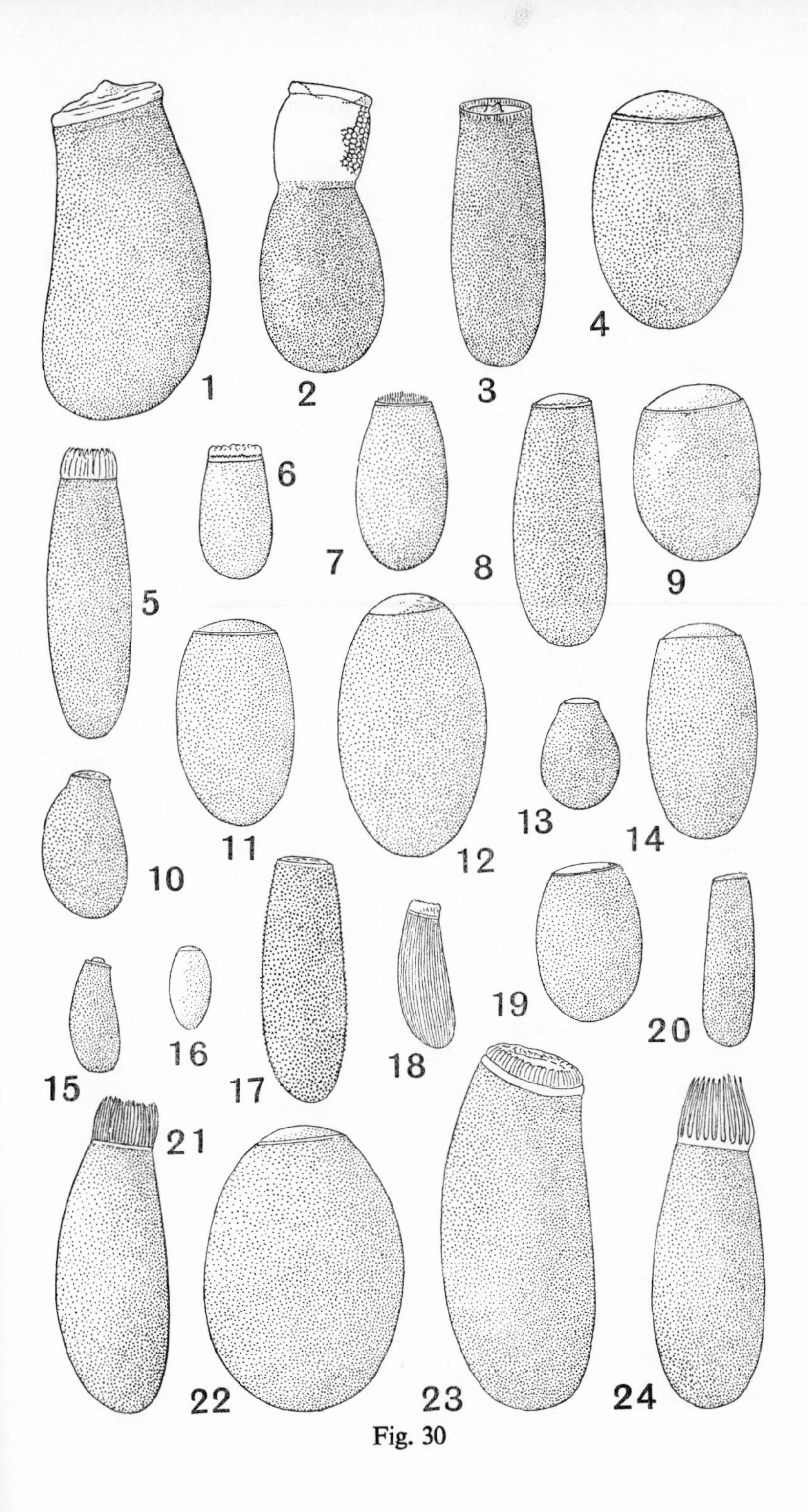

Fig. 30

females or may have been caused by enemies, such as predatory ants, driving them from their hiding places.

Distribution of Reduviidæ, particularly of diurnal species is regulated to some extent by vegetation. For example, flowering plants to which insects are attracted are often selected for the lying in wait for prey. The fact that the Reduviid in such a position is not in any way concealed, does not reduce the chances of other insects coming within reach.

Reduviidæ have been recorded from altitudes varying from sea level to approximately 7,000 feet above sea level.

Artificial light has a great attraction for both sexes of some nocturnal species. Females seem to be attracted less frequently, and, it appears, only before or after oviposition, and not during the period of development of the ova. Diurnal species have occasionally been attracted to light, but this happens as a rule when the light shines on the plant on which they are resting.

Reduviidæ kill their prey by injecting saliva at the time of piercing the body of the victim with their stylets. The saliva has an almost instant paralizing effect, except in the case of large Arthropods such as millepedes which are more resistant.

Scent glands are present in most Reduviidæ. In the adult the ostioles of these glands are located in the metasternal coxal cavities and close to the rim of the posterior acetabula. In certain genera there are two pairs of glands, the first pair in the position just mentioned and the other pair laterally near the posterior margin of the metapleural epimeron.

The position and number of the ostioles of the dorsal abdominal glands in the neanides varies, there being usually a pair on each of

---

Plate III (*facing*)

Stenocephalidæ, Lygæidæ, Pyrrhocoridæ, Largidæ, Aradidæ, Reduviidæ

1. *Lygæus elegans* Wolff 1802. Lygæidæ-Lygæinæ.
2. *Microspilus proximus* Dallas 1852. Lygæidæ-Lygæinæ.
3. *Dicranocephalus agilis* (Scopoli) 1763. Stenocephalidæ.
4. *Cænocoris floridulus* Distant 1918. Lygæidæ-Lygæinæ.
5. *Narbo fasciatus* Distant 1901. Lygæidæ-Megalonotinæ.
6. *Myodocha intermedia* (Distant) 1882. Lygæidæ-Megalonotinæ.
7. *Lohita grandis* (Gray) 1832. Largidæ.
8. *Euryopthalmus subligatus* Distant 1882. Largidæ.
9. *Melamphaus faber* (Fabricius) 1782. Pyrrhocoridæ.
10. *Mezira membranacea* (Fabricius) 1803. Aradidæ-Mezirinæ.
11. *Dysodius lunatus* (Fabricius) 1794. Aradidæ-Mezirinæ.
12. *Bagauda lucifugus* McAtee and Malloch 1926. Reduviidæ-Emesinæ.
13. *Polytoxus pallescens* Distant 1903. Reduviidæ-Saicinæ.
14. *Tribelocephala* sp. Reduviidæ-Tribelocephalinæ.
15. *Sminthus fuscipennis* Stål 1874. Reduviidæ-Reduviinæ.
16. *Tiarodes cruentus* Stål 1870. Reduviidæ-Reduviinæ.

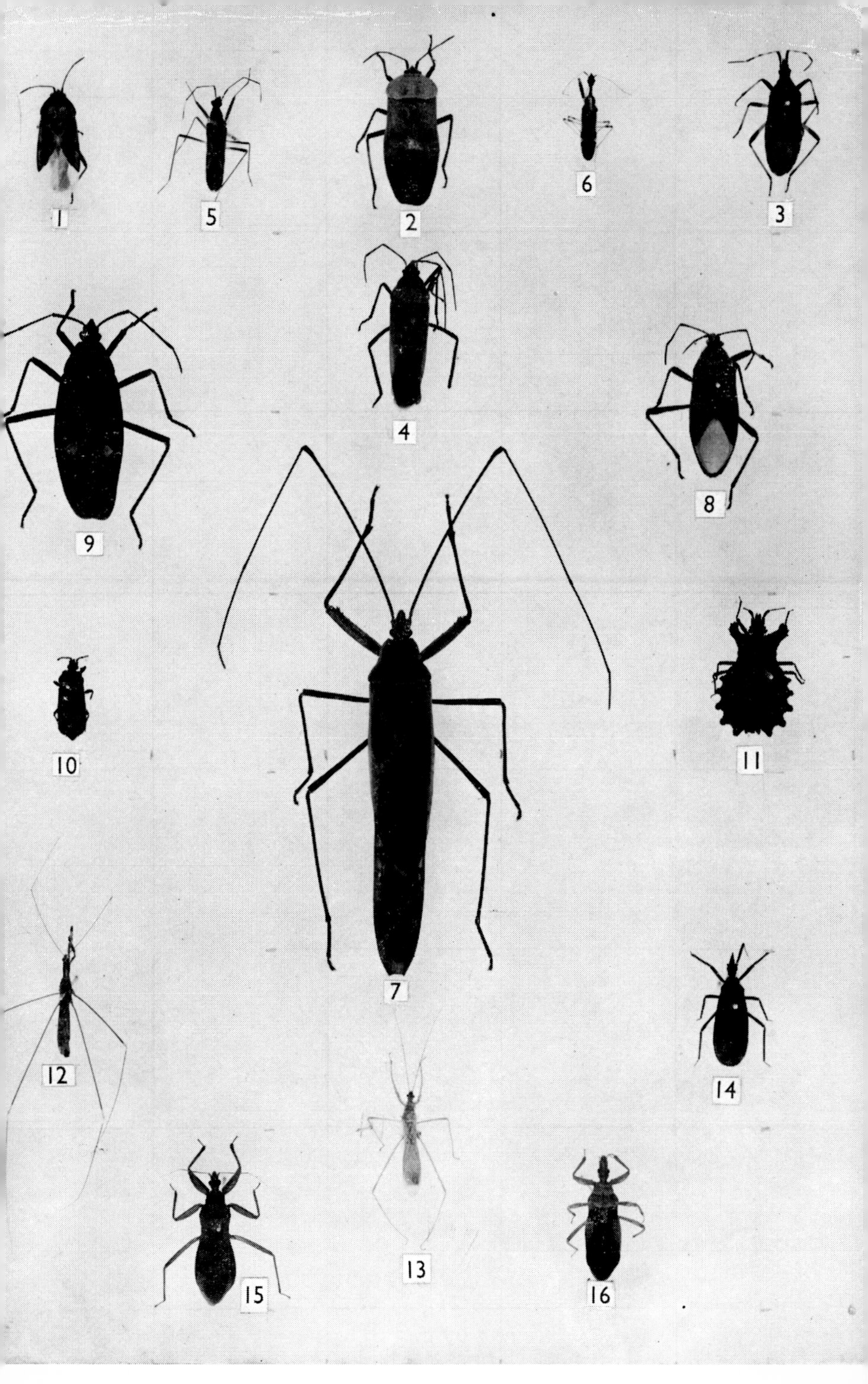

Plate III

the segments three to six or only one pair on segment five (*Toxopeusiana* Miller 1954). Several genera are provided with glandular setæ through which a glutinous substance is secreted. The setæ are mostly on the body and legs.

The Reduviidæ are divided into thirty-one subfamilies: **Holoptilinæ** (Amyot and Serville) 1843, **Emesinæ** (Amyot and Serville) 1843, **Saicinæ** (Stål) 1859, **Visayanocorinæ** Miller 1952, **Tribelocephalinæ** (Stål) 1866, **Bactrodinæ** (Stål) 1866, **Stenopodinæ** (Amyot and Serville) 1843, **Salyavatinæ** (Amyot and Serville) 1843, **Eupheninæ** Miller 1955, **Cetherinæ** Jeannel 1919, **Sphaeridopinæ** (Amyot and Serville) 1843, **Manangocorinæ** Miller 1954, **Physoderinæ** Miller 1954, **Centrocneminæ** Miller 1956, **Chryxinæ** Champion 1898, **Vesciinæ** Fracker and Bruner 1924, **Reduviinæ** (Amyot and Serville) 1843, **Triatominæ** Jeannel 1919, **Piratinæ** (Stål) 1859, **Phimophorinæ** Handlirsch 1897, **Mendanocorinæ** Miller 1956, **Hammacerinæ** (Stål) 1859, **Ectrichodiinæ** (Amyot and Serville) 1843, **Rhaphidosomatinæ** Jeannel 1919, **Harpactorinæ** (Amyot and Serville) 1843, **Apiomerinæ** (Amyot and Serville) 1843, **Ectinoderinæ** (Stål) 1866, **Phonolibinæ** Miller 1952, **Perissorhynchinæ** Miller 1952, **Tegeinæ** Villiers 1948, **Diaspidiinæ** Miller 1959.

The **Holoptilinæ** are mainly brown or light brown in colour with darker brown or piceous spots and suffusion on the hemelytra or, in *Ptilocnemus* Westwood 1840, *Rudbeckocoris* Miller 1956, *Holoptiloides* Miller 1956, the hemelytra are hyaline with large or small brown spots and suffusion. The main characters of the **Holoptilinæ** are the ample membrane of the hemelytra the corium of which is much reduced, the considerably reduced metathoracic wings, scutellum semicircular with the margin unspined but sometimes fringed, antennæ, legs and abdomen with abundant serrate and plumose setæ, trichome (to which reference is made further on) usually present on basal ventral abdominal segments, head transverse, lateral margins of pronotum dilated and dorso-ventrally compressed. Distribution Ethiopian and Oriental Regions and Australia.

In 1963, a representative of this subfamily was recorded from British Guiana. Assuming there is no confusion as the labelling, this is the first time that a Holoptiline has been recorded from the New World.

Although sometimes seen on the wing, their flight being moth-like, they are more often seen in shady places on branches or among leaves.

Information regarding the habits of the **Holoptilinæ** is scanty, consisting almost solely of observations on *Ptilocerus ochraceus* Montandon 1907 by Jacobson, who relates how he found this insect in very large numbers, adults as well as neanides, in company with numerous small black ants *Dolichoderus bituberculatus* Mayr, a soft-bodied species.

Jacobson suggested that this ant is particularly fond of sweet substances, and went on to state that 'most of the ants which I found in the above-named locality near the bugs appeared to be in a more or less paralyzed state, and the ground beneath was in some places covered an inch thick with dead ants'. Later on he was able to observe the behaviour of this Reduviid, some of which he caged with the ants. 'The bugs,' he stated, 'had fasted for about a week, the only thing I had given them being pure water sprinkled in their cage and which they readily absorbed. They were, however, none the worse for the fasting, only a few of the many hundreds I had captured having died.'

Ptilocerus Gray 1831 and other genera of this family possess what is termed a trichome. This is formed by a strong elevation or gibbosity on the third ventral segment of the abdomen which, incidentally, is fused with the second segment. This elevation is surmounted and flanked by a dense tuft of setæ and on its anterior surface is situated the ostiole of a gland which extends posteriorly along the wall of the abdomen to the fifth segment in *Ptilocnemus lemur* Westwood 1840. From the gland a substance agreeable to ants is said to be excreted and it is also considered to have intoxicating properties.

Jacobson described how a bug, on the approach of an ant raised itself on its legs so that the trichome was exposed. The ant then licked the trichome and pulled the tuft of setæ with its mandibles, but it was not until the fluid from it began to take effect that the bug inserted its mouthparts into a soft part of the ant's body. Jacobson said that sometimes the ants licked the trichome and left without being seized by the bug. Soon after, however, they were affected and thus many more ants died as a result of imbibing the fluid and not by direct attack by *Ptilocerus.*

Since these Reduviids, like other members of the family, inject their saliva—which is very potent—into the body of the victim, the possession of a subsidiary method for overcoming it seems superfluous. Much fuller investigation is desirable before the behaviour of *Ptilocerus* as described is confirmed beyond all reasonable doubt.

Jacobson also stated that the 'nymphs and adults of the bug act in exactly the same manner to lure the ants to their destruction after having rendered them helpless by treating them to a tempting delicacy'.

In the fourth instar neanide of *Ptilocerus ochraceus* there is a certain degree of gibbosity on the third abdominal segment, but whether this has an ostiole connected to a functional gland can be determined only by dissection of fresh specimens. In *Montandoniola* Villiers 1946 there is no visible trichome in the adults.

The ova of *P. ochraceus* are deposited in irregular groups in concealed places, e.g., on the inner surface of a bamboo, and are more or less covered with a white exudation. According to Kirkaldy the ovum 'is obtusely flask-shaped, flattened down ventrally; the lid is provided with a small knob. Colour brown; chorion well chitinized with finely reticulated surface composed of hexagonal and pentagonal areas. Size 1·2×0·5 mm.' An ovum of *P. lemur* dissected from a dried specimen was dark brown in colour with a whitish operculum, and ovate, constricted at the upper margin of the chorion which was glabrous. The size, 1·60 mm. In view of the small size of the adult, this is a large ovum. Probably only six to eight develop at one time (Fig. 28, 4).

No trichome appears to be present in the neanides of *P. lemur*.

Emesinæ are small or moderately large, slender, fragile insects with raptorial anterior legs and very long, slender median and posterior legs. The anterior tarsi are sometimes composed of three segments of variable length or of one segment with one or two claws. The median and posterior tarsi are very short and composed of three segments. The anterior coxæ are long. The eyes are variable, but usually prominent and with relatively large and few facets. The ocelli are absent. The antennæ are long and slender, sometimes longer than the body and occasionally with abundant sericeous setæ.

Both alate and apterous forms occur and the presence or absence of wings is not related to sex, and when these appendages are absent, the thorax may exhibit modifications but not the sexual organs. There are no odoriferous glands.

The hemelytra are long and narrow with a narrow corium which may extend along the greater part of the costal margin and even to the apex of the membrane.

The **Emesinæ** are found in many different habitats, in particular in those places which are humid and poorly lighted. Some species have been found in caves, but they do not exhibit adaptations to a cavernicolous existence. Species which have been collected in caves include, *Myiophanes fluitaria* McAtee and Malloch 1926, *Bagauda lucifugus* McAtee and Malloch 1926 and *Bagauda cavernicola* Paiva 1919, the first two collected in Malaya, the last-mentioned collected in Assam.

A stridulatory furrow is present, so far as is known, in all genera, but is somewhat variable in structure. In some genera it is apparently not striate, e.g., *Guithera* Distant 1906; in others the striations are extremely feeble. Generally the striæ are relatively few in number, coarse and widely separated. In *Stenolæmus plumosus* Stål 1871 and *S. crassirostris* Stål 1871, the striæ become progressively coarser towards the posterior end of the furrow; in *S. marshalli* Distant 1903, *S. decarloi* Wygodzinsky 1947 and *S. bogdanovi* Oshanin 1870, the furrow is coarsely striate in the posterior half only. The furrow is similarly striate in *Eugubinus reticolus* Distant 1915.

Not a great deal is known concerning the ecology of the **Emesinæ**. Their food consists mainly of small insects such as gnats and other small flies which are often present in the places frequented by them. The **Emesinæ** are nocturnal and are frequently recorded as having been attracted to artificial light.

Those ova which have been examined are mostly cylindrical, somewhat narrow and may have chorionic processes or a long filament arising from the operculum. The chorion is generally smooth and may be ribbed or have short scales or spines, as in *Stenolæmus marshalli* (Fig. 28, 2).

The **Emesinæ** are distributed in all zoogeographical Regions.

Visayanocorinæ (Fig. 31). In this subfamily there are two genera, *Visayanocoris* Miller 1952 with one species, *nitens*, and *Carayonia* Villiers 1951 with two species, *culiciformis* Usinger 1952 and *camerunensis* Villiers 1951.

They are characterized by a smooth, shining integument, head without a transverse sulcus on the vertex, by having no ocelli, the scutellum with a long, slender apical spine, the anterior tibiæ with a flattened, acute projection on the inner surface apically, the pronotum almost without spines on the postero-lateral angles of the posterior lobe, stridulatory furrow present.

V. nitens occurs in the Philippine Islands, *C. culiciformis* in Ceylon and *camerunensis* in West Africa. Apparently their habitat is the canopy of forest trees, but apart from this nothing is known about their ecology.

Saicinæ are mostly fragile, elongate insects with long, slender legs and antennæ, the anterior coxæ opening straight downwards, the lower surface of the head with spines, the rostrum with bristles, the head with a distinct transverse sulcus, the pronotum and scutellum and postscutellum with spines usually. The eyes are prominent and with large facets; ocelli are absent.

The body is minutely pilose or sericeous. Odoriferous glands are apparently absent; a stridulatory furrow is present.

Very little is known about their habits and development. The ovum of an unidentified species of *Polytoxus* Spinola 1852 which I have examined is cylindrical, with one side feebly curved and with a flat operculum (Fig. 28, 3).

The ovum of *Polytoxus marianensis* Usinger 1946 is oblong oval, broadly rounded on one side and scarcely rounded on the other. It is rounded posteriorly and has the micropylar end carinate around a relatively small operculum. The chorion is glabrous and white. The total length is 0·75 mm.

The **Saicinæ** are entirely nocturnal and are distributed in the Old and New Worlds.

Bactrodinæ are small, slender insects with the hemelytra having only one large cell in the membrane, the anterior femora spined, the lower anterior margin of the prothorax produced distinctly beyond the upper margin in the middle of which the head is inserted, the third rostral segment as long or a little longer than the second segment, body usually slender and the anterior legs usually raptorial. A stridulatory furrow present.

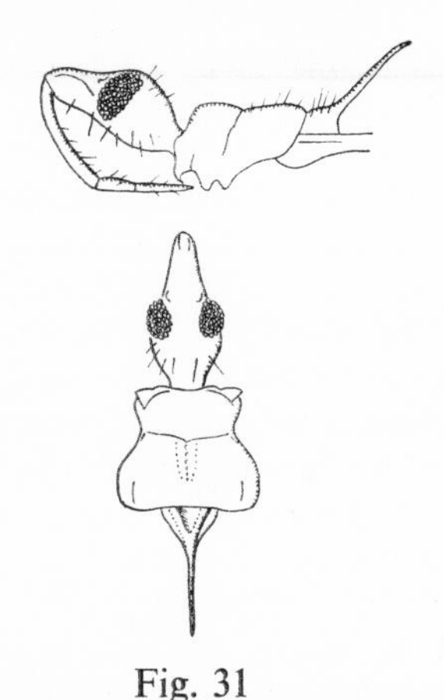

Fig. 31

Visayanocoris nitens Miller 1952 (Reduviidæ-Visayanocorinæ).

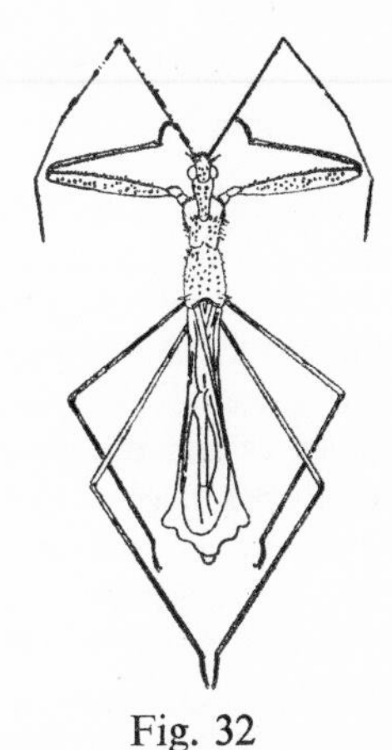

Fig. 32

Bactrodes spinulosus Stål 1862 (Reduviidæ-Bactrodinæ).

The **Bactrodinæ** (Fig. 32), are apparently allied to the **Emesinæ.** There is no ecological information available. They are apparently nocturnal, but have been reported as having been seen on plants in the daytime. Distribution Neotropical Region.

Stenopodinæ. Mostly somewhat narrow, small to large Reduviids of dull colouration, piceous or testaceous, many species having large black or piceous spots on the membrane of the hemelytra. They are characterized by the long antennæ, the usually elongate cylindrical head with a distinct neck, presence of inter-antennal spines and by the venation of the hemelytra which generally have a large discal cell. This type of venation is, however, not constant, the cell being absent and the venation of the corium markedly different in *Anacanthesancus* Miller 1955, *Canthesancus* Amyot and Serville 1843 and *Thodelmus* Stål 1859. A stridulatory furrow is present and is variable in structure, e.g. in *Aulacogenia* Stål 1870, it is coarsely striate posteriorly, but the striations on the anterior part are visible only under high magnification.

In some genera, for example, *Oncocephalus* Klug 1830, *Sastrapada* Amyot and Serville 1843, the anterior femora are thick and have rows of spines on the lower surface, while in *Padasastra* Villiers 1948, *Staccia* Stål 1865 and *Neostaccia* Miller 1940 moderately long spines are present. A *fossula spongiosa* is absent from all genera except *Canthesancus* and *Anacanthesancus*. Some members are strongly spinose, for example, *Echinocoris* Miller 1949 and *Parechinocoris* Miller 1949.

The **Stenopodinæ** are nocturnal in habits and are often attracted to artificial light. So far as is known, females oviposit in the soil. Ova (Fig. 28) which have been examined are ovate, with the operculum feebly convex, or are sub-spherical with short chorionic processes.

With regard to their being attracted to artificial light, although both sexes are attracted, it seems that the females are not susceptible until after oviposition has taken place.

Dorsal glands are present on the fourth and fifth segments of the abdomen in neanides.

The prey of **Stenopodinæ** has apparently not been recorded apart from one instance, namely *Aulacogenia cheesmanæ* Miller 1952 found feeding on a neanide of *Parabryocoropsis typicus* China and Carvalho 1951 (Miridæ) in New Britain.

Distribution world-wide, the Old World having by far the larger number of genera.

Tribelocephalinæ are dull-coloured, mostly dark brown, black or fulvous insects with dense tomentose clothing on head, body and corium, which conceals to a great extent the actual form of these parts, the head in particular. Fine and moderately long setæ are usually interspersed among the tomentosity, and in some species similar setæ arise from between the facets of the eyes, which are not at all prominent, but definitely flattened so that they do not project beyond the lateral margins of the postocular region of the head and are sometimes almost contiguous dorsally; the ocelli are

---

Fig. 33 (*facing*)
Neanides of Reduviidæ

1. *Petalochirus umbrosus* Herrich-Schaeffer 1853 (Salyavatinæ). 5th instar.
2. *Centrocnemis* sp. (Reduviinæ). 4th instar.
3. *Cimbus* sp. (Ectrichodiinæ). 4th instar.
4. *Cleontes genitus* Distant 1903. (Diaspidiinæ). 5th instar.

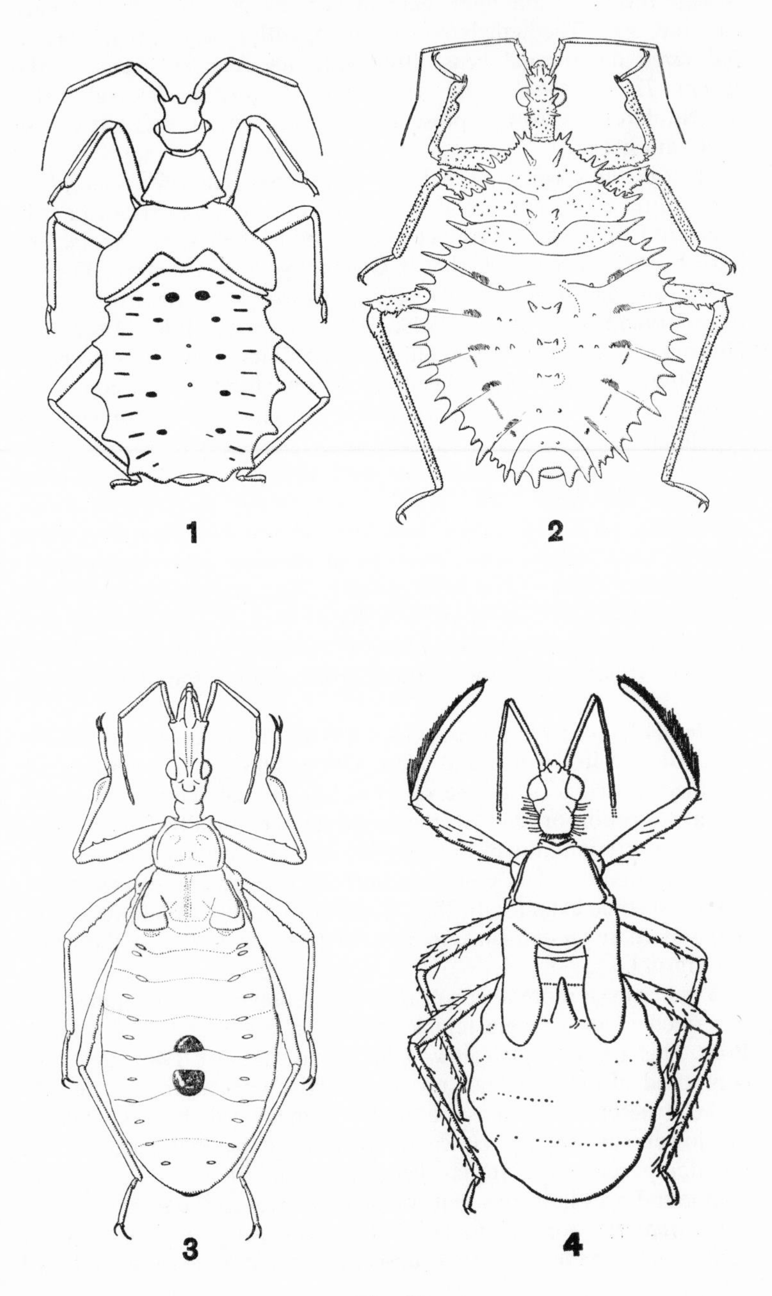

Fig. 33

absent; the basal antennal segment is thickened and much longer than the head. The hemelytra are ample, with a very narrow corium and large membranal cells. Brachypterous females occur in the genera *Tribelocephala* Stål 1853 and *Afrodecius* Jeannel 1919. A stridulatory furrow is present, the striæ being somewhat coarse in certain genera.

Most of the species are included in the two genera *Tribelocephala* and *Opisthoplatys* Westwood 1834, but there are some extraordinarily aberrant forms which have been described in the last fifty years. The most bizarre of these are the six Oriental genera *Apocaucus* Distant 1909, *Megapocaucus* Miller 1954, *Gastrogyrus* Bergroth 1921, *Acanthorhinocoris* Miller 1940, *Matangocoris* Miller 1940 and *Homognetus* Bergroth 1923, all of which differ from *Tribelocephala* and *Opistoplatys* among other things, in having the tarsi composed of two segments and in the particular arrangement of the tomentosity on the head.

Apocaucus and *Megapocaucus* are remarkable for the unusual shape of the head, the vertex being greatly elevated laterally. Abundant, very long curved setæ, more or less fused for the greater part of their length, arise from these elevated areas and conceal the dorsal surface of the head entirely. Another character of these genera is the very considerably reduced corium. So far, only two species are known, *Apocaucus laneus* Distant 1909 and *Megapocaucus laticeps* Miller 1954, the former from India, the latter from Java.

Afrodecius, of which there are six species known up to the present, is distributed in Central and West Africa. This genus is remarkable for the unusual form of the rostrum, the second segment of which has a projection or spur on the inner surface apically. The purpose of this spur is unknown, but possibly it may have some connection with stridulation. The colour pattern of one species of *Afrodecius* agrees to some extent with that of certain Lycidæ (Coleoptera), and it has been suggested that, in view of this, the habits of the genus are diurnal.

A genus even more aberrant than those previously mentioned was described from a single female found on the island of Fernando Poo. This genus, *Xenocaucus* China and Usinger 1949, has tarsi composed of one segment, no wings and no eyes. Another feature is the structure of the basal antennal segment which is concave on the lower surface, the concavity forming a resting place for the remaining segments. It has been suggested that the absence of compound eyes indicates that *Xenocaucus* lives in the soil.

Opistoplatys appears to be a composite genus which could be legitimately divided into two sub-genera, the difference being based on the structure of the head and rostrum mainly, and also the type and abundance of tomentose clothing. Female *Tribelocephala* and *Opistoplatys* apparently are not active until after oviposition, that is, if it is justifiable to infer this from the fact that the females in collections which have been examined have all been found to have completely empty abdomens.

Hardly anything is known of the habits of the **Tribelocephalinæ,** but in the main, they appear to be nocturnal. They are often attracted to artificial light. Occasionally they are found among vegetable debris.

The **Tribelocephalinæ** are distributed in the warm regions of the Old World with the exception of Madagascar.

Salyavatinæ are dull-coloured insects with black or pale yellow spots, particularly on the connexivum, in some species. The basal antennal segment is usually moderately thick, the antennophores prominent, the head and pronotum have long spines sometimes and the rostrum is thick and short. A *fossula spongiosa* is present on the anterior and median tibiæ. In *Petalochirus* Palisot Beauvois 1805, *Alvilla* Stål 1874 and *Syberna* Stål 1874, the anterior tibiæ are compressed and expanded. In all genera the anterior tarsi are composed of two segments. The scutellum sometimes has basal lateral spines or tubercles as well as an apical spine. A stridulatory furrow is present. The postero-lateral angles of each connexival segment is usually acute or dentate.

The **Salyavatinæ** are mostly nocturnal, but some species of *Petalochirus* are often found on herbage in the daytime in the open. They are frequently attracted to artificial light.

There appear to be three different types of ovum in this subfamily, the strangest being those of *Petalochirus umbrosus* Herrich Schaeffer 1853 and *Elaphocranum* Bergroth 1904 (Fig. 28, 9). These resemble a cone flattened on three sides, and with the operculum, which is consequently triangular, also flat. The pole opposite the operculum is also somewhat flattened. The ova of *Petalochirus brachialis* Stål 1858 (Fig. 28, 12) and of *P. obesus* Miller 1940 and of a species of *Alvilla* Stål 1874 (Fig. 28, 13) are of another type, being more or less spherical with the operculum pyriform in outline. The ova of some species of *Lisarda* Stål 1859 (Fig. 28, 10), are sub-ampulliform and somewhat compressed near the apex which results in the operculum being pyriform in outline, but it is feebly convex with a cylindrical elevation.

A fifth instar neanide of *P. umbrosus* is shown in Figure 33, 1.

It is not unusual for Reduviidæ, when handled, to attempt to pierce the fingers (often successfully) with the stylets. *P. umbrosus*, however, in my experience, has shown no readiness to bite, the only reaction tot his kind of treatment being vigorous stridulation and the emission of fluid from the metathoracic glands.

The ostioles of the dorsal abdominal glands in the neanides of *P. umbrosus* and *Lisarda rhodesiensis* Miller 1950 are located at the base of the fourth, fifth and sixth segments.

Little is known of the food preferences of the **Salyavatinæ.** *Lisarda* spp. have been noticed attacking worker termites and also the alate forms in the teak forests of Java.

With the exception of one genus, *Salyavata* Amyot and Serville 1843, the **Salyavatinæ** are distributed in the tropical and sub-tropical regions of the Old World. *Salyavata* is distributed in the Neotropical Region.

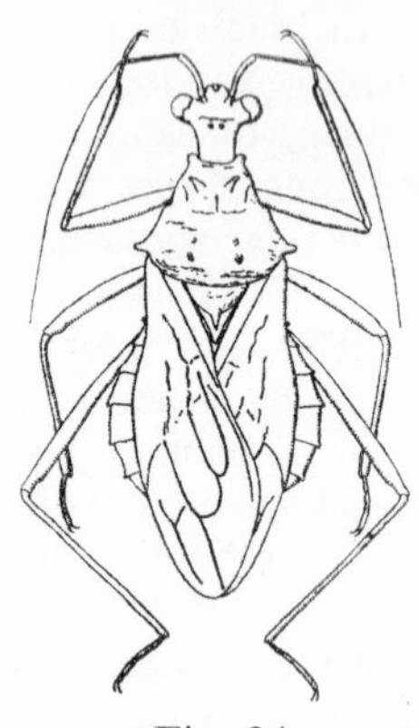

Fig. 34

Eupheno pallens (Laporte) 1832 (Reduviidæ-Eupheninæ).

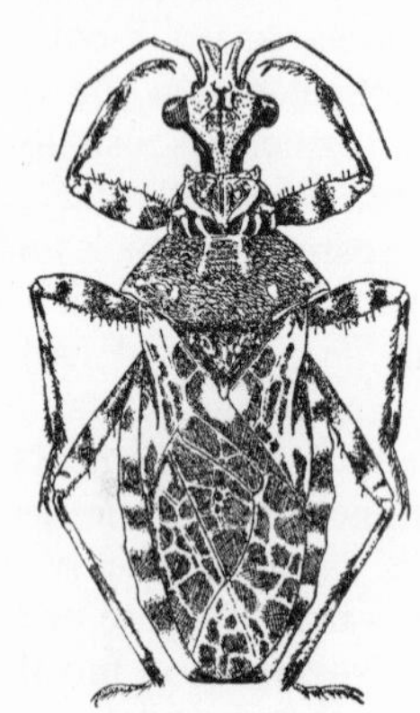

Fig. 35

Cethera marmorata Miller 1950 (Reduviidæ-Cetherinæ).

Eupheninæ (Fig. 34), up to the present time, contains one genus, *Eupheno* Gistel 1848 in which there are three species, *pallens* (Laporte) 1832, *histrionicus* Stål 1862 and *rhabdophorus* Breddin 1898. The subfamily characters are: head transverse, eyes pedunculate, ocelli not elevated, anteocular with a bifurcate process, juga produced, pronotum with spines and tubercles, scutellum with a spine arising behind the apex, prosternum produced anteriorly, segment six of abdomen obsolete midventrally, segment seven very wide and angulate, segment eight visible, metathoracic glands with two pairs of ostioles, one each in the metasternal depression and adjacent to the coxal cavity and one on each metapleural epimeron, on a lamellar elevation, anterior and median tibiæ with a *fossula spongiosa*, apophyses of female genital segments strongly developed.

The ovum of an unidentified species of *Eupheno* is cylindrical, rather narrow and with a flat operculum (Fig. 30, 20).

Nothing is known, apparently, about the ecology of the species referred to. The structure of the female genitalia suggests that the ova are inserted into some substance or other. Distribution, Neotropical.

Cetherinæ (Fig. 35). Members of this subfamily are small, active insects characterized by the more or less pedunculate eyes, elevated ocelli, vertex lobately produced, tuberculate and spined pronotum, scutellum with an apical spine and sometimes with basal lateral spines, habitus somewhat depressed dorso-ventrally.

The venation of the corium has the following characters: R+M, M+Cu very short; IA meeting Cu at about the middle and continuing to the base of the internal cell of the membrane; IA with a branch vein arising near the apex of the internal cell which is considerably narrower than the external cell. In addition to the pair of metathoracic gland ostioles located in the metasternal depression and adjacent to the inner margin of the acetabulum there is another pair of ostioles. These are on the metapleural epimeron and near its posterior margin.

There are four genera known up to the present time. They are *Cethera* Amyot and Serville 1843, *Cetheromma* Jeannel 1917, *Caridomma* Bergroth 1894, *Carcinomma* Bergroth 1894 and *Caprothecera* Breddin 1903.

The ova of *Carcinomma simile* Horvath 1914 (Fig. 28, 23), are subspherical with minute chorionic filaments and those of *Cethera musiva* (Germar) 1837, ovate with the anterior end somewhat constricted. Neanides of *Cethera* spp. have been found under logs and stones. The food of both adults and neanides is probably termites.

The **Cetherinæ** are confined to the Ethiopian Region and Madagascar.

Manangocorinæ (Fig. 37). This subfamily which contains one species, *Manangocoris horridus* Miller 1954, is characterized by the transverse head with the anteocular strongly declivous, almost vertical, small and narrowly separated ocelli, the tuberculate basal antennal segment, the transverse pronotal lobes in which the anterior lobe is shorter than the posterior lobe, the scutellum with a long apical spine, the membranous hemelytra, the anterior tibiæ incrassate and laterally compressed apically and with no *fossula spongiosa* and by abundantly setose posterior tibiæ, tarsi very long.

Nothing is known of its habits or its developmental stages. Distribution, Borneo.

Sphaeridopinæ. The characters of representatives of this subfamily are the small head with large eyes narrowly separated at their lower surfaces, a rostrum with a very short and thick basal segment, slender straight second, and a very short third segment. In the genus *Volesus* Champion 1899 articulation between the second and the third segment is very difficult to define. The median and posterior coxæ are widely separated. Another character is the long stridulatory furrow which is striate, however, for a little more than half its length posteriorly.

The genera belonging to this subfamily, *Sphaeridops* Amyot and Serville 1843, and *Volesus* are Neotropical. The genus *Eurylochus* Torre Bueno 1914, should not be placed in this subfamily. Nothing is known about their habits or food.

Physoderinæ (Fig. 38). Rather small, dull-coloured Reduviidæ with the following characters: head and body tuberculate, the tubercles low and bearing setæ, usually spatulate; head elongate with a transverse sulcus behind the eyes; ocelli present; rostrum straight with the second segment much longer than segments one and three together; both lobes of pronotum more or less transverse; scutellum with the apex produced, the produced portion flattened and sulcate; hemelytra complete with the first anal vein forming part of the internal cell of the membrane extended to about the middle of the claval suture and also produced towards the apex of the membrane; vein Sc not coalescing with R; M not connected with R; R+M diverging at apical margin of the corium; membrane extended backwards along costal margin of corium; anterior and median femora spined on lower surface; *fossula spongiosa* absent, stridulatory furrow present.

The metathoracic wings in representatives of this subfamily are coloured and may be entirely infumate, entirely yellow or infumate with the basal half yellow.

Fig. 36 (*facing*)
Ova of Reduviidæ

1. *Tydides rufus* Serville 1831. 3·00 mm. Piratinæ.
2. *Santosia maculata* (Fabricius) 1781. 2·20 mm. Ectrichodiinæ.
3. *Philodoxus principalis* (Distant) 1903. 3·20 mm. Ectrichodiinæ.
4. *Hammacerus cinctipes* (Stål) 1858. 3·00 mm. Hammacerinæ.
5. *Cleontes ugandensis* Distant 1912. 3·30 mm. Diaspidiinæ.
6. *Beharus cylindripes* (Fabricius) 1803. 5·00 mm. Apiomerinæ.
7. *Tapirocoris limbatus* Miller 1954. 1·90 mm. Harpactorinæ.
8. *Apiomerus lanipes* (Fabricius) 1803. 3·00 mm. Apiomerinæ.
9. *Rhaphidosoma maximum* Miller 1950. 4·50 mm. Rhaphidosomatinæ.
10. *Graptoclopius lieftincki* Miller 1954. 3·00 mm. Harpactorinæ.
11. *Acanthiscium* sp. 1·60 mm. Harpactorinæ.
12. *Phonolibes tricolor* Bergroth 1912. 2·70 mm. Phonolibinæ.
13. *Lopodytes nigrescens* Miller 1950. 3·00 mm. Rhaphidosomatinæ.
14. *Korinchocoris insolitus* Miller 1941. 1·50 mm. Harpactorinæ.
15. *Endochus* sp. 2·00 mm. Harpactorinæ.
16. *Sinea undulata* Uhler 1894. 1·40 mm. Harpactorinæ.
17. *Hæmatoloecha* sp. 1·90 mm. Ectrichodiinæ.

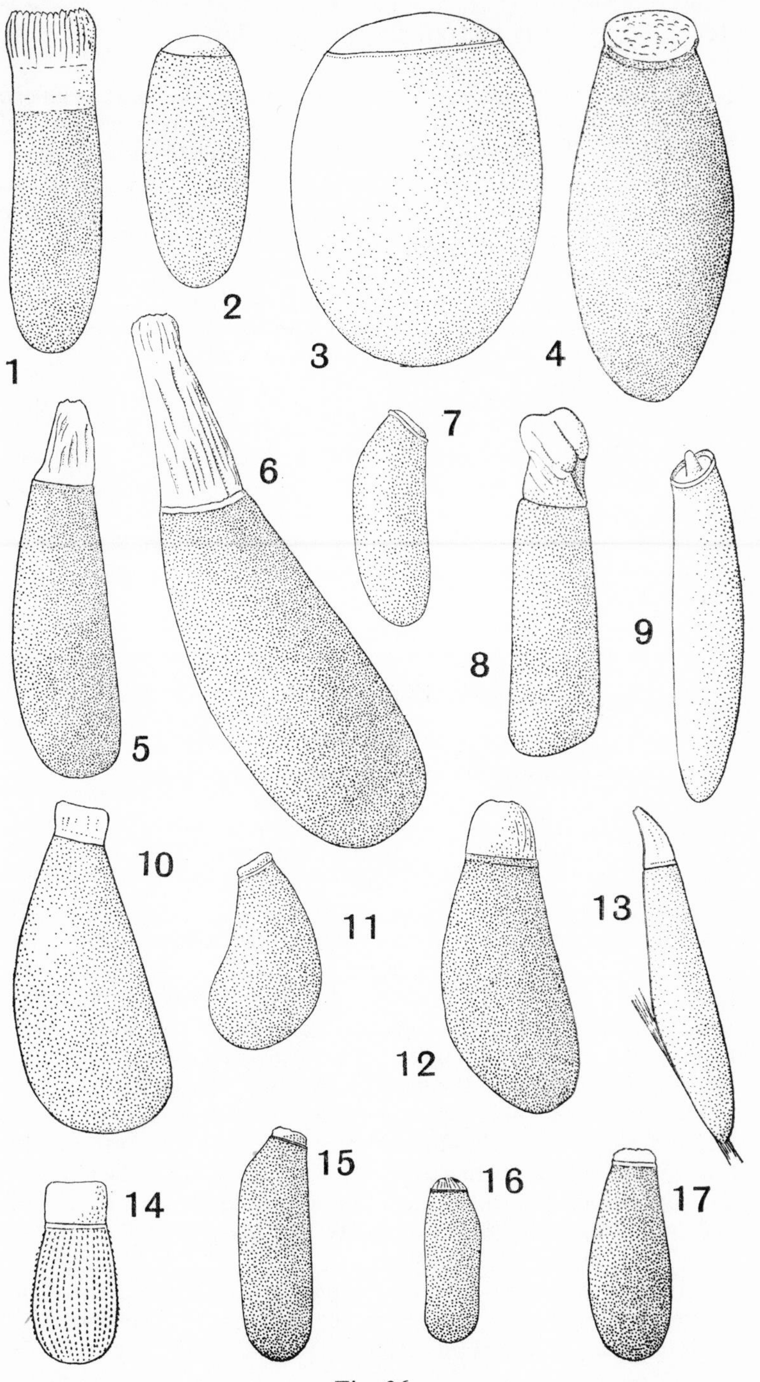

Fig. 36

There are three genera known at the present time, namely, *Physoderes* Westwood 1844, *Neophysoderes* Miller and *Physoderoides* Miller 1956, the first-named distributed in the Oriental Region, Madagascar and Mauritius, the last two recorded from Madagascar.

Physoderes spp. are found mainly among decaying vegetable matter. One species, *P. curculionis* China 1935 is said to prey on coleopterous larvæ in decaying banana stems.

The ovum of *Physoderes patagiata* Miller 1941 (Fig. 30, 15) is cylindrical with one side somewhat shorter than the other and

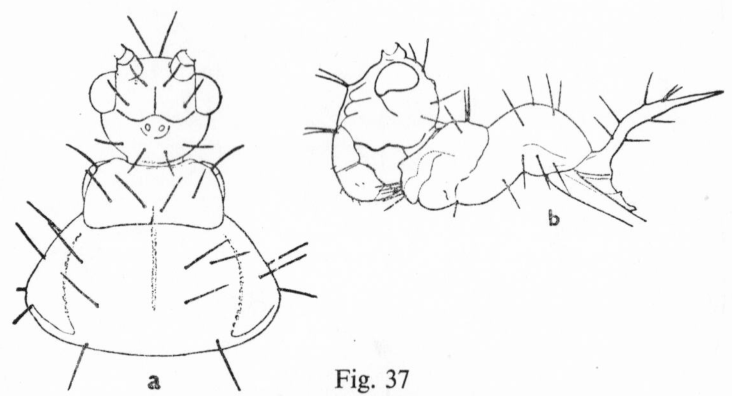

Fig. 37
Manangocoris horridus Miller 1954 (Manangocorinæ).

straight. The operculum has, in its centre, a truncate, rounded, cylindrical elevation and the differentiated portion of the chorion is narrow. In colour it is brownish-yellow and the operculum and differentiated portion of the chorion whitish.

Centrocneminæ. This subfamily includes the genera *Centrocnemis* Signoret 1852, distributed in Northern India and Malaysia, *Neo-centrocnemis* Miller 1956, from Malaysia, Celebes, Philippine Islands, Indo-China and Formosa, *Paracentrocnemis* Miller 1956, from South India and Ceylon and *Centrocnemoides* Miller 1956 recorded from the Malay Peninsula, Sumatra and Java.

They are all dull-coloured, greyish insects with dark brown spots and suffusion on body and hemelytra. They have the rostrum with four visible segments, the basal segment distinct, the head, body and legs spined and tuberculate, metathoracic glands paired with one ostiole in each metacoxal cavity and another at the external angle of the metasternal epimeron.

Not much is known of their ecology beyond the fact that their habitat is the trunks of trees in jungle country. The somewhat dorso-ventrally compressed habitus of both adults and neanides suggests that they may spend some of their life under the loose bark of trees.

The ova of some species have been studied after dissection from dried specimens. They are smooth, ampulliform with the opercular end oblique and with the upper surface of the operculum with irregular ridges or a stellate pattern of ridges.

The ovum of *N. signoreti* (Stål) 1863 is shown in Fig. 30, 1 and the fourth instar neanide of a *Centrocnemis* species in Fig. 33, 2.

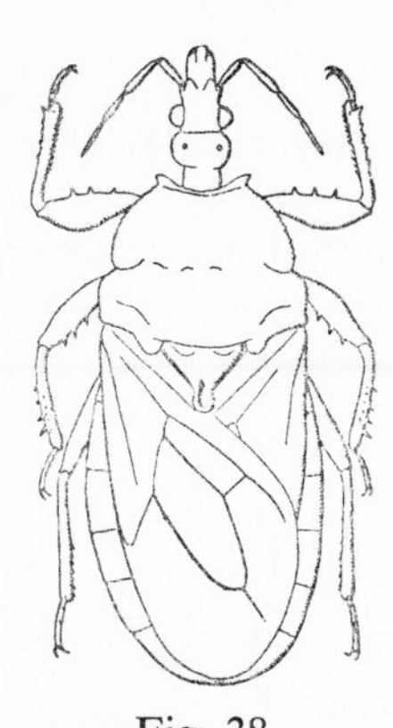

Fig. 38

Physoderes patagiata Miller 1941 (Physoderinæ).

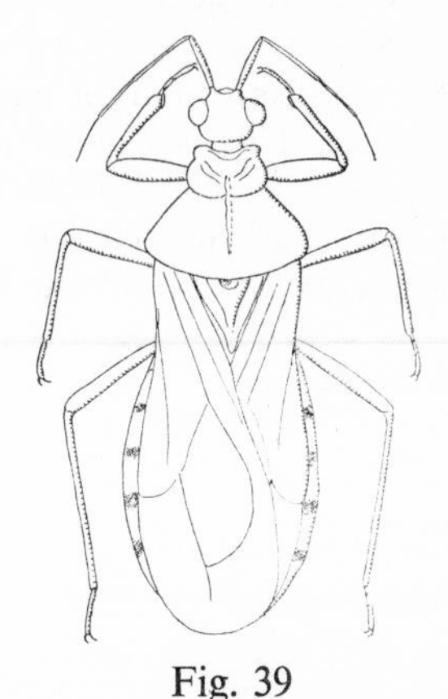

Fig. 39

Chryxus tomentosus Champion 1898 (Reduviidæ-Chryxinæ).

Chryxinæ (Fig. 39). Small or medium sized insects covered with bristle-like hairs. Head much wider and higher than long and strongly declivous anteriorly; sulcus on vertex distinct; ocelli absent or present—in the latter event, situated between the eyes; rostrum robust and short, strongly curved; basal segment of antennæ short and robust, the remaining segments more slender; transverse sulcus on pronotum situated nearer the apical margin than the posterior margin; legs simple; tarsi with three segments; hemelytra with a distinct corium; membrane with one cell; abdominal glands in neanides present on the third, fourth and fifth segments; stridulatory furrow present.

This subfamily contains three species, *Chryxus tomentosus* Champion, 1898, *Wygodzinskyiella travassosi* (Lent and Wygodzinsky) 1944, the former from Panama, the latter from Brazil and *Lentia corcovadensis* Wygodzinsky 1946.

Nothing appears to be known about the ecology of these species.

Vesciinæ (Fig. 40). This subfamily contains small dark-coloured insects characterized mainly by the short anteocular portion of the head; the presence or absence of ocelli; eyes with large facets; anterior femora incrassate; anterior tibiæ curved apically; tarsi long and slender. The hemelytra may be fully developed or brachypterous with two cells in the membrane. The anterior pronotal lobe is considerably longer than the posterior lobe and the scutellum is produced apically. A stridulatory furrow is present.

The genera so far known are *Vescia* Stål 1865, *Pessoaia* Costa Lima 1941, *Microvescia* Wygodzinsky 1943, *Mirambulus* Breddin 1901, all Neotropical, *Chopardita* Villiers 1948 from West Africa and the Sudan.

There is no information regarding their habits.

Reduviinæ. This is a large and composite subfamily represented in all zoogeographical Regions. It contains many genera, all of which are characterized by the absence of a discal cell in the hemelytra, by the scutellum with an apical spine or tubercle and

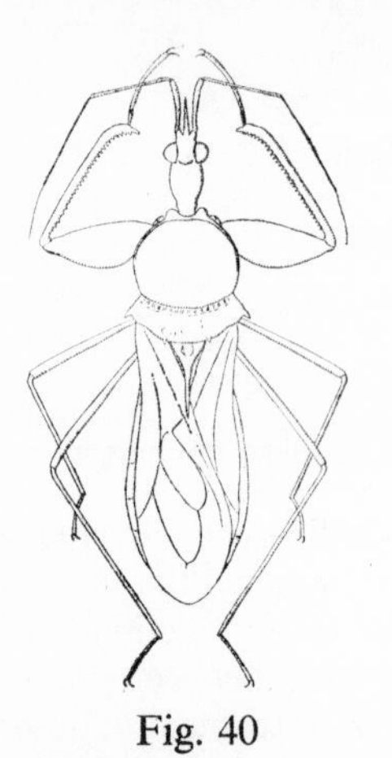

Fig. 40
Vescia adamanta Haviland 1931 (Reduviidæ-Vesciinæ).

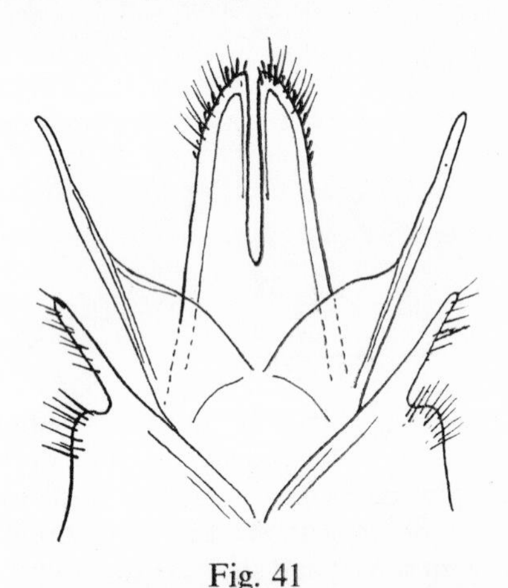

Fig. 41
Velitra rubropicta Amyot and Serville 1843 (Reduviidæ-Reduviinæ) (female genitalia).

sometimes latero-basal spines. The ocelli are well-developed except in apterous forms in which they are greatly reduced or absent. The pronotum has a transverse sulcus situated near the middle of the segment. The legs are more or less slender and the tarsi have three segments. Odoriferous glands are usually present in the neanides and are located on the fourth, fifth and sixth segments. Genera without such glands are *Gerbelius* Distant 1903 and *Durganda* Amyot and Serville 1843. A *fossula spongiosa* is present on both the anterior and median tibiæ in most genera. It is absent from *Psophis* Stål 1863, *Cheronea* Stål 1863, *Nalata* Stål 1858, *Microlestria*

Stål 1872, *Euvonymus* Distant 1904, and is present on the anterior tibiæ only of *Voconia* Stål 1865, *Stachyogenys* Stål 1870, *Gerbelius* Distant 1903, *Haplonotocoris* Miller 1940, *Pasira* Stål 1859, *Pasiropsis* Reuter 1881, *Nannolestes* Bergroth 1913, *Heteropinus* Breddin 1903 and *Durganda*. In *Croscius* Stål 1874 it is extremely small and is present on both anterior and median tibiæ. Males and females of *Holotrichius* Burmeister 1835 have no *fossula spongiosa* on either anterior or median tibiæ, except the species *insularis* Distant 1903, which has this organ on the median tibiæ only.

A stridulatory furrow is present in most genera, both in the adults and in the neanides of a more advanced stage. It is, however, absent from *Psophis* and *Euvonymus*. In *Staliastes* Kirkaldy 1900 the striæ have partly disappeared, e.g., in *S. rufus* Laporte 1832 and *S. zonatus* (Walker) 1873, the furrow has no striæ posteriorly. The striæ are absent from *S. malayanus* Miller 1940 and from *Heteropinus*.

The **Reduviinæ** comprise mostly alate forms but brachypterous forms also occur. In certain genera, namely, *Edocla* Stål 1859, *Paredocla* Jeannel 1914 and *Holotrichius*, the females are always apterous, but the males may be fully alate or apterous. In *Diplosiacanthia* Breddin 1903 the females are apterous and both sexes of *Ectmetacanthus* Reuter 1882 are brachypterous. Apterous males and females as well as alate males occur in *Psophis*, but so far as is known, alate females have not been discovered. Brachypterous and micropterous individuals occur in *Acanthaspis* Amyot and Serville 1843.

As previously stated, the **Reduviinæ** is a composite subfamily containing several genera for which new subfamilies should be made. Among these may be mentioned the genera with a strongly flattened habitus, a reduced or no stridulatory furrow and widely separated coxæ, to mention a few of the distinguishing characters: for example, *Durganda*, *Staliastes*, *Apechtia* Reuter 1881, *Apechtiella* Miller 1948, *Sminthus* Stål 1865, *Velitra* Stål 1865, *Heteropinus* Breddin 1903, all of which pass a part of their lives under the loose bark of dead trees. The females have the apophyses of the genital segments strongly developed forming an ovipositor suitable for the insertion of ova into crevices in decaying bark or wood (Fig. 41).

The habits of **Reduviinæ** are various. Some genera live in human dwellings, in stables or fowl runs—for example, species of *Reduvius* Lamarck 1801—some under the loose bark of dead trees, some in desert or sub-desert areas, namely *Reduvius* (some spp.), *Parthocoris* Miller 1950 and *Holotrichius; Khafra* Distant 1902 has been found in caves but is not a true cavernicolous insect; *Alleocranum biannulipes* Montrouzier and Signoret 1861, a small and strongly setose species, is often found in stored products such as rice, on the insect pests of which it feeds. Its mode of life has brought about its almost cosmopolitan distribution. *Acanthaspis petax* Stål 1866 has been found in termite nests in Uganda.

The habit of accumulating debris on the body and legs is met with in the genera *Reduvius*, *Acanthaspis* and *Paredocla* and is common both to adults and neanides. The debris which may consist of soil fragments, remains of insect prey or of vegetable matter, adheres to the secretory hairs from which a glutinous substance flows. It is still a complete mystery what purpose this masking of the body with debris serves, since the genera which have the habit live almost entirely in concealment in secluded and dark places. The opinion is expressed sometimes that the disguise is an aid in the capture of prey and also that it is protective against enemies. Nevertheless, the accumulating of debris is deliberate and not fortuitous, as may be proved by removing it and then allowing the denuded bug access to similar material. Within a few minutes it will restore the mantle of debris by casting it on the body with its legs.

Reduviinæ appear to be general feeders and will attack most Arthropods within certain size limits. There is little information of species confining themselves to one sort of prey. *Phonergates bicoloripes* Stål 1855 is said to prey on ticks and *Reduvius personatus* Linnæus 1758 to feed on the bedbug, but it is not known whether ticks and bedbugs respectively are the sole kinds of food sought by these two Reduviids.

Platymerus rhadamanthus Gerstaecker 1873, a large black species with red markings, distributed in eastern Africa has been reported as preying on adults of the coconut beetle, *Oryctes monoceros* Oliver in Zanzibar. The saliva secreted by both the neanides and adults of this species is highly virulent. This fact was impressed on me on one occasion on a collecting trip in the Uluguru Mountains in Tanzania (formerly Tanganyika Territory). On this occasion one of my assistants brought me a specimen of *P. rhadamanthus* which he held in a cleft stick. Not paying the attention I should have, I grasped it with a finger and thumb to remove it to the collecting jar. Immediately the Reduviid forced its stylets into the top of my thumb and injected saliva. This caused excruciating pain which extended along the arm and across the chest and persisted for more than twenty-four hours.

Reduviinæ deposit their ova usually without a glutinous covering in the soil, among vegetable debris or in crevices in the bark of trees. The ova known up to the present are mainly ovoid or cylindrical, with a smooth chorion and feebly convex operculum. Exceptions are the ova of *Psophis consanguinea* Distant 1903 (Fig. 30, 2) which have a large differentiated portion which is also reticulate with the upper margins deflected, the ova of *Pasiropsis vidua* Miller 1954 (Fig. 28, 21) which are cylindrical with long chorionic filaments, the ova of *Durganda rubra* Amyot and Serville 1843 (Fig. 30, 10), subampulliform. Other cylindrical ova are those of *Sminthus* spp. (Fig. 30, 5), *Dyakocoris vulnerans* (Stål), 1863 (Fig. 30, 2), *Velitra alboplagiata* Stål 1859 (Fig. 30, 5), and of *Veltra rubropicta* Amyot and Serville 1843 (Fig. 30, 17), the last-mentioned with a minutely granulose chorion, and those of *Sminthus* spp. have moderately long chorionic fllaments. The ovum of *Dyakocoris* has a more or less flat operculum with an irregularly truncate median elevation and short differentiated portion. The ovum of *Korinchocoris insolitus* Miller 1941, has a rather long differentiated portion similar to that of the ova of some **Harpactorinæ.** It has also minute, curved, prominent setæ on the chorion.

Triatominæ (Plate IV). This subfamily contains the most important of all the Reduviidæ. Its members are exclusively predaceous on mammalian and avian blood and are vectors in the Neotropical Region of a malady known as Chagas disease (or American trypanosomiasis) which has affects on human beings similar to those of sleeping sickness, a scourge in some areas of tropical Africa.

The **Triatominæ** are characterized by the elongate head, straight rostrum, complete or partial absence of the transverse sulcus on the vertex and the absence of neanidal abdominal glands. In the adults, according to Brindley, the "stink gland passes laterally into a groove which runs along the internal face of the emarginated edge of the metacoxal cavity . . . and communicates with the exterior by a small pore."

A stridulatory furrow is usually present.

The **Triatominæ** are mostly sombre-coloured insects with sometimes red or yellow markings. *Triatoma rubrofasciata* (de Geer) is tropicopolitan, *T. migrans* Breddin 1903 and *T. pallidula* Miller 1941 have been recorded from Malaysia and *T. novaguineæ* Miller 1958 from New Guinea.

Another species found outside America is the Indian *Linshcosteus carnifex* Distant 1904. The affinities of this species to other **Triatominæ** are problematic; it has no *fossula spongiosa* and the rostrum does not extend beyond the posterior margin of the eyes, consequently there is no stridulatory furrow.

The ova of **Triatominæ** have in many cases been described and figured by Usinger. They are mostly ovate with a smooth chorion and a more or less flat operculum, and are deposited loosely in places where the females congregate.

Triatoma rubrofasciata, it has been alleged, is responsible in some measure for the propagation of Kala Azar or oriental sore. This unpleasant condition and Chagas Disease are not the result of trypanosomes or other organisms being injected when the bug is feeding. The disease-provoking organisms are present in the fæces of the bug and when the patient scratches or rubs the spot at which irritation has been set up by the bite, he introduces them into the blood-stream. The relation of **Triatominæ** to mammals and birds is more fully dealt with in the chapter 'Heteroptera Associated with Mammals and Birds'.

Under laboratory conditions, several species of **Triatominæ** have been recorded as feeding on lizards and snakes. There is a possibility that lizards may become hosts for *Trypanosoma cruzi*, since they live together with *Triatoma* and their host the woodrats (*Neotoma*).

Hammacerinæ (Plate IV). This subfamily contains two genera, *Microtomus* Illiger 1807 and *Homalocoris* Perty 1833, both of which exhibit a remarkable type of antennal structure not to be found in any other Heteroptera. The first antennal segment is short and robust and the second segment is divided into many segments from eight to forty. Other characters are the prominent eyes and ocelli which are located between them. A stridulatory furrow is present. Dorsal abdominal glands of the neanides are on the third, fourth and fifth segments.

Little is known about the ecology of this subfamily apart from the fact that the habitat is under loose bark of decaying trees. The ovum of *Hammacerus cinctipes* Stål 1858 is elliptical with a flat operculum. (Fig. 36, 4).

The **Hammacerinæ** comprise mostly dark-coloured insects with whitish stripes or yellowish spots and are confined to South America, Mexico and the southern parts of the United States of America.

Piratinæ (Plate IV). Mostly black and piceous Reduviids. Some species are partly yellow and red with parts of the legs similarly coloured. The principal characters are the somewhat elongate head, usually smooth and shining, the longer than wide pronotum which is often either elongately striate or without sculpturation and with a constriction well lbehind the middle. The anterior and median tibiæ have a *fossula spongiosa*, which, in some genera, for example *Ectomocoris* Mayr 1865, may extend almost the entire length of the tibiæ. The anterior coxæ are often longer than wide and flattened on the outer surface. Incidentally the *fossula spongiosa*, in some species, extends beyond the apex of the anterior tibiæ. Alary polymorphism occurs in some genera, apterous specimens being often met with. Dorsal glands are present in the neanides on segments three, four and five.

Piratinæ, so far as is known, are mostly nocturnal. In the daytime they live in secluded places under stones and in crevices. They are often attracted to artificial light. It would appear the ova are deposited by the females mainly in the soil. Those that have been examined are ovate with a smooth chorion and with fairly long chorionic processes which expand outwards on eclosion. The ovum of *Catamiarus brevipennis* (Serville) 1831 is subampulliform with the chorion glabrous and with filaments curved inwards apically on the apical margin; the operculum is concave with a cylindrical elevation constricted medially on the upper surface and the margin with filaments curved inwards.

The ovum of *Pteromalestes nyassæ* (Distant) 1877 is cylindrical, subampulliform with moderately long chorionic processes curved inwards somewhat, apically; the chorion is dull greyish olivaceous, the processes dark grey; the apical margin of the chorion narrowly black. At the time of eclosion of the neanide the processes expand outwards. The ovum is deposited by the female in the soil.

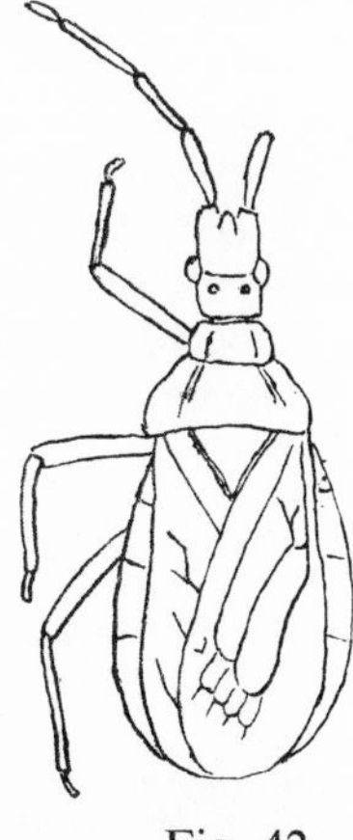

Fig. 42

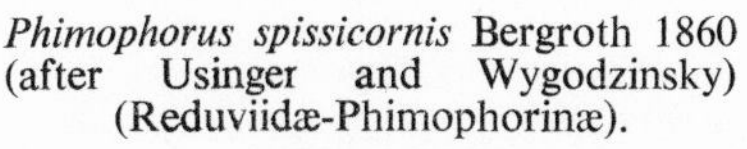

Phimophorus spissicornis Bergroth 1860 (after Usinger and Wygodzinsky) (Reduviidæ-Phimophorinæ).

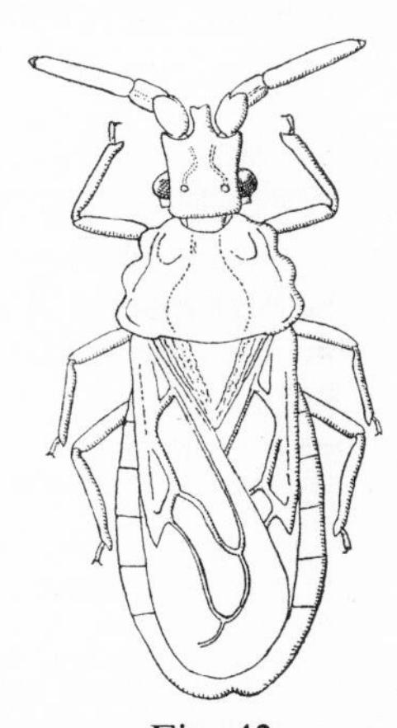

Fig. 43

Mendanocoris browni Miller 1956 (Reduviidæ-Mendanocorinæ).

This species is, apparently, not entirely nocturnal, one specimen, a female having been found in Southern Rhodesia in full sunshine on a grass stem at about three feet from the ground.

Piratinæ appear to be general feeders. *Fusius rubricosus* (Stål) 1855, an Ethiopian species, is able to inflict considerable pain by its bite. In addition to the pain caused by the injected saliva there is also a danger that the wound will become septic on account of other matter being introduced at the same time.

I was once bitten by a female of this species in Southern Rhodesia, which I had found in fresh cowdung in which it was probably seeking coleopterous and dipterous larvæ.

In this particular instance the site of the puncture became septic shortly afterwards, and, although treated, the affected area remained in a suppurating condition for some days.

The **Piratinæ** are widely distributed in the New and Old Worlds.

Phimophorinæ (Fig. 42). A subfamily containing a single genus and species, *Phimophorus spissicornis* Bergroth 1886 (Fig. 40). This is a small Reduviid with somewhat flattened habitus. The body is granulose and has the longer setæ serrulate. The head is moderately elongate with spiniferous tubercules on each side of the gular region and with the dorsal surface granulose. The antennæ are short and robust, particularly the basal segment which is sub-rectangular. The eyes and ocelli are small. Rostrum straight with the basal segment much longer than segments two and three together. Legs short and granulose, the tibiæ laterally compressed with a short *fossula spongiosa.* Hemelytra somewhat wide with the membrane containing three cells. A stridulatory furrow is present.

Phimophorus is a somewhat aberrant Reduviid and was in the first instance placed by Bergroth in the Aradidæ. Handlirsch (1897) considered it to be a member of the **Phymatinæ** (Phymatidæ), but subsequently believed it to be allied to *Aulacogenia* Stål 1870, a genus of the **Stenopodinæ**. Its systematic position has been discussed by Carayon, Usinger and Wygodzinsky, but no definite conclusion was reached.

Plate IV (*facing*)
Reduviidæ

1. *Holoptilus agnellus* Westwood 1874. Holoptilinæ.
2. *Stenopoda* sp. Stenopodinæ.
3. *Neocentrocnemis signoreti* (Stål) 1863. Centrocneminæ.
4. *Sphæridops amoenus* Lepeletier and Serville 1825. Sphæridopinæ.
5. *Platymerus guttatipennis* Stål 1865. Reduviinæ.
6. *Zelurus spinidorsis* (Gray) 1832. Reduviinæ.
7. *Triatoma rubrovaria* (Blanchard) 1843. Triatominæ.
8. *Ectomocoris maculicrus* Fairmaire 1858. Piratinæ.
9. *Ectrichodia antennalis* (Stål) 1859. Ectrichodiinæ.
10. *Glymmatophora erythrodera* (Schaum) 1853. ♂. Ectrichodiinæ.
11. *Glymmatophora erythrodera* (Schaum) 1853. ♀. Ectrichodiinæ.
12. *Hammacerus cinctipes* (Stål) 1858. Hammacerinæ.
13. *Phemius tibialis* Westwood 1837. Harpactorinæ.
14. *Rhinocoris neavei* Bergroth 1912. Harpactorinæ.
15. *Panthous ectinoderoides* Bergroth 1913 Harpactorinæ.
16. *Ectinoderus bipunctatus* Amyot and Serville 1843. Ectinoderinæ.
17. *Apiomerus vexillarius* Champion 1899. Apiomerinæ.
18. *Diaspidius scapha* (Drury) 1782. Diaspidiinæ.
19. *Phonolibes tricolor* Bergroth 1912. Phonolibinæ.
20. *Tegea femoralis* Stål 1870. Tegeinæ.
21. *Rhaphidosoma maximum* Miller 1950. Rhaphidosominæ.

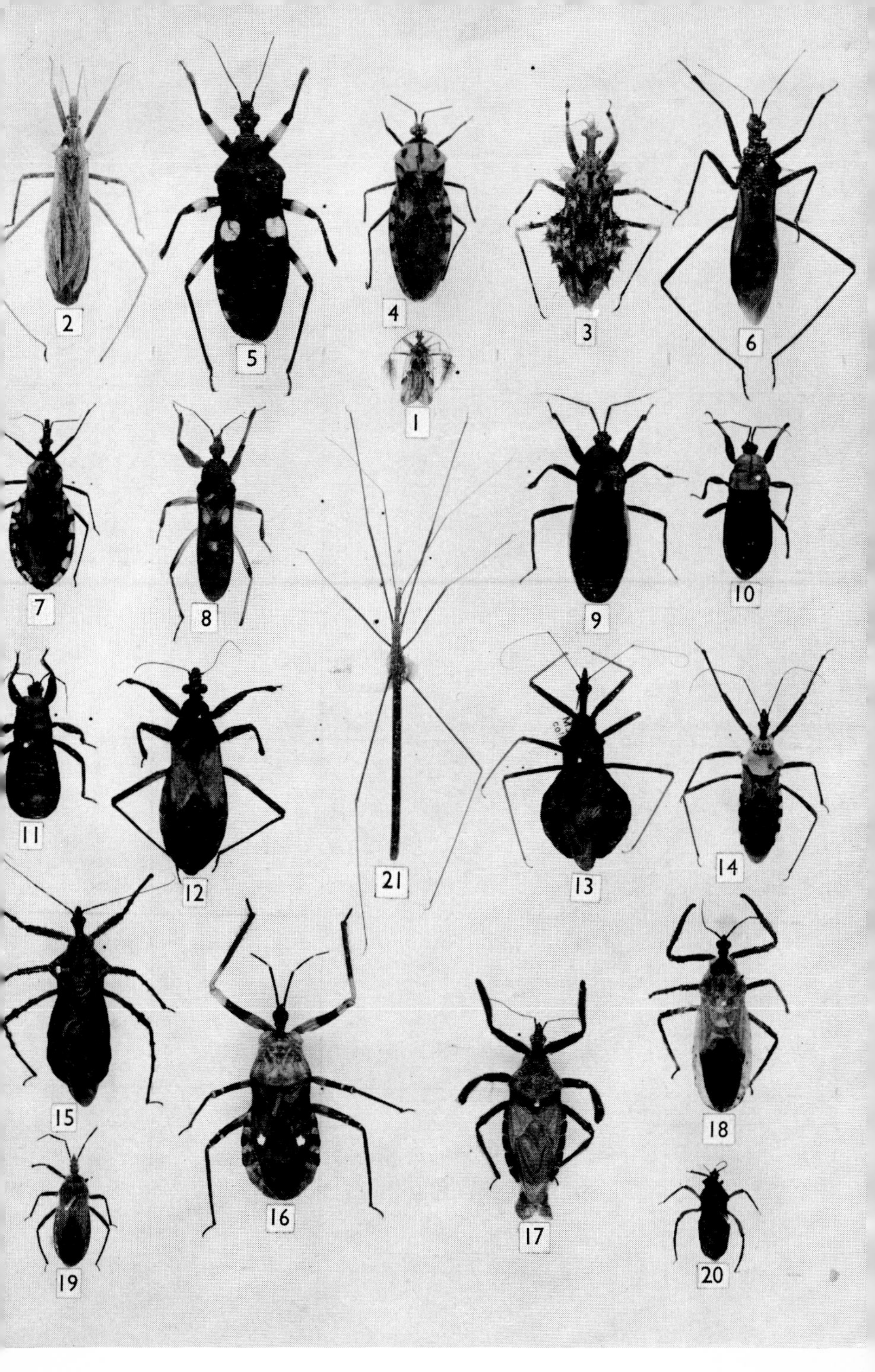

Plate IV

Handlirsch remarked, as regards its phylogenetical relationship, *Phimophorus* is indeed off the greatest interest, since it indicates, perhaps, the path by which the predatory and certainly younger Phymatidæ have branched of from the other phytophagous bark-dwellers, the Aradidæ.

Nothing is known of the ecology of this insect which comes from the Neotropical Region.

Mendanocorinæ (Fig. 43). This subfamily contains, up to the present, one genus with two species, *Mendanocoris browni* Miller 1956 and *M. milleri* Usinger and Wygodzinsky 1964. The former was found in the Solomon Islands in coconut palm and the latter in Penang, Malaysia, without details of its habitat being recorded. These small and somewhat remarkable Reduviids, the size being under five millimetres, which have a superficial resemblance to certain Aradidæ on account of the flattened habitus, the quadrate head, the short rostrum, which is partly concealed by the bucculæ when not extended, the widely separated median and posterior coxæ, appear to be related to *Phimophorus* Bergroth 1886 **(Phimophorinæ).** Other characters comprise, antennæ thick, with the apical segment very short and partly concealed when viewed from above, in the apex of the third segment: pronotum with a transverse sulcus much nearer to the anterior margin than to the posterior margin; scutellum produced apically; tibiæ laterally compressed and with a *fossula spongiosa;* tarsi with two very short segments; lateral margins of gula region expanded. *M. browni* was discovered in the axils of younger fronds of coconut palms on the inside and outside of the 'felt'. This is all that is known about its ecology.Ecological details regarding *M. milleri* are also lacking.

Ectrichodiinæ (Plate IV). A moderately large subfamily represented in the tropics of the Old and New Worlds. Its members are mostly robust with a thick integument which may be glabrous or strongly sculptured, particularly the dorsal surface of the abdomen of apterous forms. Many are brightly coloured, mainly red and yellow but many also are entirely black or piceous with a violaceous or greenish metallic tinge. Some have the abdomen entirely metallic green.

The ventral surface of the abdomen is often transversely carinulate intersegmentally. The hemelytra are complete, but the division between the corium and membrane is often indistinct. A stridulatory furrow is present and is also sometimes well-developed in the later instars of the neanides. The rostrum is thick and curved in most genera, but is straight in *Cimbus* Hahn 1831, *Katanga* Schouteden 1903, *Afrocastra* Breddin 1903 and *Katangana* Miller 1954. A rostrum differing very considerably from the curved and straight types is that of *Xenorhyncocoris caraboides* Miller 1938 from Sumatra and *X. princeps* 1948 from Malaya. In *princeps* it is thick, strongly curved at the extreme base of the basal segment which is also somewhat compressed dorso-ventrally. The second segment is more or less normal, that is to say, cylindrical, but the external surface is flattened. The third segment is somewhat compressed laterally. In *caraboides* the basal segment is sinuate and strongly compressed dorsoventrally, the second segment is strongly swollen and feebly laterally compressed, the third segment somewhat compressed laterally. It is also of note that the prosternum in these two species is strongly produced posteriorly (beyond the base of the mesosternum) consequently the stridulatory area is very long.

The number of antennal segments in **Ectrichodiinæ** varies from four to eight, but no known genus, however, has five segments.

The scutellum is more or less quadrate with two or three apical spines, the lateral ones usually widely separated except in a few genera.

The legs, particularly the anterior and median femora, are robust and may have spines on the lower surface. In some species the anterior femora and trochanters have a raised shagreened area on the lower surface. This affords an additional aid in the gripping powers of the *fossula spongiosa* which are present on the anterior tibiæ of most genera.

Odoriferous glands are present, but their position and number on the dorsal surface of the abdomen of neanides are not constant. For example, in *Philodoxus* Horvath 1914 and *Toxopeusiana* Miller 1954 there is only one on the fifth segment; in *Cimbus* there is a gland on the fourth and fifth segments; in *Marænaspis* Karsch 1892 it is difficult at first sight to locate the position of the glands since the ostiole on segment four is very small, and at the base of segment six is a sclerotized plate scarcely separated from the dorsal plate on segment five. This is a much larger plate with the ostioles transverse. Usually the ostioles are at the base of the fifth and sixth segments. In *Paravilius* Miller 1955 and *Xenorhyncocoris*, *Glymmatophora* Stål 1853 and *Ectrichodia distincta* Signoret 1858 the ostioles are similarly located. The kind of odour of the fluid from the neanides and adults does not appear to have been noted with the exception of that from *Marenaspis corallinus* Miller 1950 which resembles that of verbena.

The prey of the majority of the genera is unknown. Some genera, for example, *Ectrichodia* Lepeletier and Serville 1825, *Marænaspis*, *Scadra* Stål 1859, appear to be restricted in their choice of food, however, and feed only on Myriapoda (*Spirostreptus* spp. *et al.*). This is a curious fact, seeing that the body fluids of these Arthropoda have corrosive properties able to raise blisters on the human skin.

The **Ectrichodiinæ** are mainly nocturnal but may occasionally be seen during the daytime in bright sunshine. Apterous forms, e.g. *Glymmatophora*, sometimes leave their usual habitat under logs or stones and move about in the open. These excursions are not always voluntary but may be provoked by an invasion of predatory ants.

Alary polymorphism is frequent, the reduction or absence of wings being confined mainly to the females, but both alate and apterous males as well as apterous females have been recorded, for example, in the genus *Marænaspis*. Genera in which the apterous condition occurs include *Glymmatophora*, *Mindarus* Stål 1859, *Xenorhyncocoris*, *Distirogaster* Horvath 1914, *Hæmatorrhopus* Stål 1874, *Katanga*, *Afrocastra*, *Katangana*. Brachypterous forms appear in the genus *Ectrychotes* Burmeister 1835.

Oviposition takes place in the soil, the ova being placed in a loose mass without the addition of any glutinous substance. A species of *Scadra* in Malaya envelops the ovum in a whitish gelatinous substance which subsequently hardens. The ova of **Ectrichodiinæ,** so far as is known, are sub-spherical, cylindrical, ovate or elliptical without sculpturation and with a feebly convex operculum.

Rhaphidosomatinæ (Plate IV). The characters of this subfamily are: a slender, elongate body, long slender antennæ and legs, cylindrical head, small eyes and ocelli (the ocelli may be absent), and a straight rostrum with the second segment the longest. A stridulatory furrow is present. In *Lopodytes* Stål 1853 and *Hoffmannocoris* China 1940 both sexes are alate and in *Rhaphidosoma* Amyot and Serville 1843 and *Leptodema* Carlini 1892, neither is alate.

The females oviposit on plant stems and affix their elongate, cylindrical ova singly at an angle to the substratum (Fig. 36, 9, 13).

There is little definite information regarding the prey of species belonging to this subfamily. The long straight rostrum suggests that they may seek their prey—such as lepidopterous stem-borers—living in plants. There is one record of *Hoffmannocoris chinai* feeding on *Syntomis atricornis* (Lepidoptera-Syntomidæ).

The secretion from the glands of the adults of *Rhaphidosoma circumvagans* Stål 1855 is very pungent and often reveals their presence in the herbage. The ostioles of the glands of the neanides of this species are located at the base of the fourth and fifth abdominal segments.

Members of the genera *Rhaphidosoma* and *Lopodytes*, both neanides and adults, have the habit of lying, when at rest, along a stem with their antennæ and anterior legs stretched out in front and the median and posterior legs out backwards and close to the body.

The **Rhaphidosomatinæ** are confined to the desert areas and savannahs of the Old World and are found mainly among grasses or on low bushes.

Perissorhynchinæ (Fig. 44) Miller 1952. This subfamily contains only one genus, *Perissorhynchus*, which has the following characters: rostrum straight with the basal segment very short, segments two and three almost equal in length and relatively thick, almost entirely membranous hemelytra with the corium extended nearly to the apex of the membrane; a stridulatory furrow is present. There is one species, *lloydi* Miller 1952 from West Africa. Nothing is known about its ecology.

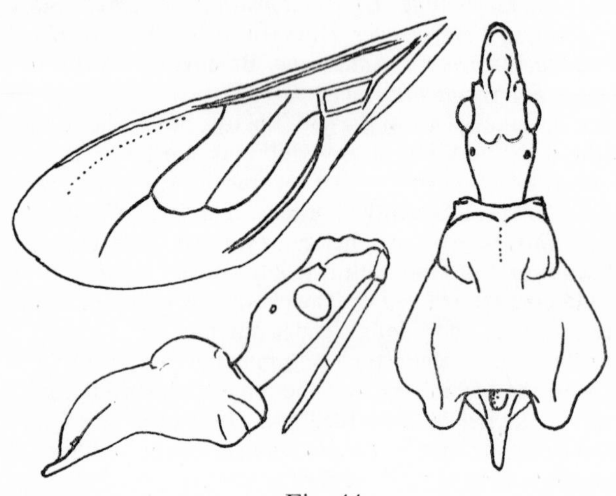

Fig. 44

Perissorhynchus lloydi Miller 1952. (Reduviidæ-Perissorhynchinæ.)

Harpactorinæ (Plate IV). All zoogeographical regions have representatives of this subfamily which contains more genera than any other subfamily of the Reduviidæ. The principal characters are: rostrum composed of three visible segments and usually curved but straight in a few genera; ocelli small and often elevated: antennæ mostly slender with four segments, but with an intercalary segment sometimes between the basal segments; one genus *Gattonocoris* Miller 1957 with the apical segment club-like; well-developed hemelytra with (in most genera) a 4-6 angled cell formed by the branching of the cubital vein; legs more or less slender, nodulose, spinose or tuberculate; secretory hairs present on the body and legs in some genera, among which *Rhinocoris* Hahn 1834, *Paramphibolus* Reuter 1887, *Peprius* Stål 1859, *Cosmolestes* Stål

1866. The hairs are present in both neanides and adults and it is rare not to find them with a good deal of vegetable and other debris adhering to them, a condition which would appear to be embarrassing rather than useful. Alary polymorphism is infrequent and is confined to the genera *Coranus* Curtis and *Dicrotelus* Erichson 1842.

A stridulatory furrow is present in most but not all genera; it is absent from the genus *Aphonocoris* Miller 1950 and in *Piestolestes* Bergroth 1912 the prosternum has an extremely narrow median sulcus, but the striæ are apparently absent.

Dorsal abdominal glands are present on the third, fourth and fifth segments and the dorsal plates are mostly quadrate or ovate in outline. Exaggerated forms of dorsal plates are to be seen in *Pantoleistes* Stål 1853, in which they are elongate and conical (Fig. 47, 3). The actual position of the ostioles of the glands can be confirmed only by preparing specimens in KOH. It is usually stated that in the **Harpactorinæ** these ostioles are on the third, fourth and fifth segments. Actually they are at the base of the fourth, fifth and sixth segments and in *Rhinocoris albopunctatus* Stål 1855, for example, the ostiole on segment four is located at some distance from the basal margin of the segment, that on segment five is less distant and the ostiole on segment six is on the basal margin. In *Coranus carbonarius* Stål 1855, *Rhinocoris segmentarius* Germar 1837 and *Phonoctonus nigrofasciatus* Stål 1855, the ostiole is at the base of the fourth, fifth and sixth segments. In *R. segmentarius* I have found the vestiges of a gland in the adult.

Many **Harpactorinæ** have irregular patches or linear elevated areas composed of a white wax-like substance on the head, body and corium. This substance may be seen in *Peprius*, *Rhinocoris*, *Aprepolestes* Stål 1868, *Acanthiscium* Amyot and Serville 1843 and several others. The function of this substance has not yet been explained, but it may possibly be connected with ecdysis.

There are many forms represented in the **Harpactorinæ.** Striking peculiarities in structure may be seen in *Arilus* Hahn 1831 (Neotropical), which has the pronotum strongly elevated with denticles on the margin of the elevation; *Notocyrtus* Burmeister 1835 (Neotropical) has the pronotum swollen and concealing the scutellum. In *Sava* Amyot and Serville 1843 (Neotropical) the pronotum is strongly gibbose and produced posteriorly to cover almost the whole abdomen. Its margins are tuberculately produced. In this genus also the prosternum is produced posteriorly beyond the anterior coxæ and is thus longer, probably, than in any other Harpactorine genus. In the Oriental genus *Panthous* Stål 1863, the pronotum is produced posteriorly concealing the scutellum.

Sphagiastes Stål 1853 has a strongly spinose pronotum with some of the spines arising from elevations. The connexival segments are strongly produced, flattened and spinose. A fifth instar neanide of *Sphagiastes ramentaceus* (Germar) 1837 is shown in Fig. 47, 1.

Many brightly coloured genera are found in this subfamily. Among these may be mentioned *Havinthus* Stål 1859 (Australia, *Vitumnus* Stål 1864, *Callilestes* Stål 1866 and *Rhinocoris* (Palæarctic Ethiopian and Oriental) of which *Rhinocoris imperialis* Stål 1859, is probably the most striking, the peculiarly hymenopteroid *Acanthiscium* and *Cosmonyttus* Stål 1867 (Neotropical), *Eulyes* Amyot and Serville 1843 (Oriental) and *Phonoctonus nigrofasciatus* Stål (Ethiopian). Incidentally *Vitumnus* exhibits a considerable range of colour variation.

Harpactorinæ are diurnal and both adults and neanides frequent flowers to which their prey is attracted, or they roam over bushes and herbage in search of lepidopterous larvæ. *Rhinocoris fuscipes* (Fabricius) 1787, an abundant and widely distributed species in South-East Asia is often found on *Polanisia viscosa* (a plant with sticky hairs) feeding on insects trapped thereon. *Cosmolestes picticeps* Stål 1859, also a common species in the same region, frequents *Passiflora fœtida*, a low plant possessing bracts with sticky hairs to which other insects become attached.

I have stated that the **Harpactorinæ** are diurnal, but there is, however, one exception, so far as I know, *Hediocoris tibialis* Stål

---

Fig. 45 (*facing*)

Ova of Reduviidæ-Harpactorinæ

1. *Helonotus versicolor* Distant 1912. 2·00 mm.
2. *Vesbius* sp. 1·10 mm.
3. *Heza sphinx* Stål 1863. 2·30 mm.
4. *Cargasdama noualhieri* Villiers 1951. 1·60 mm.
5. *Rhinocoris nitidulus* (Fabricius) 1781. 3·40 mm.
6. *Cydnocoris* sp. 2·00 mm.
7. *Macracanthopsis nodipes* Reuter 1881. 1·20 mm.
8. *Henricohahnia wahnschaffei* Breddin 1900. 2·40 mm.
9. *Paracydnocoris distinctus* Miller 1953. 2·70 mm.
10. Egg mass of *Isyndus heros* (Fabricius) 1803, showing exit hole of hymenopterous parasite.
11. *Henricohahnia vittata* Miller 1954. 2·40 mm.
12. *Villanovanus dichrous* (Stål) 1863. 2·20 mm.
13. *Rihirbus barbarus* Miller 1941. 1·60 mm.
14. *Polydidus* sp. 1·60 mm.
15. *Kibatia* sp. 1·40 mm.
16. *Repipta fuscipes* Stål 1855. 2·20 mm.
17. *Notocyrtus depressus* Stål 1872. 1·70 mm.
18. *Zavattariocoris senegambiæ* Miller 1954. 2·10 mm.
19. *Sphagiastes ramentaceus* (Germar) 1837. 3·00 mm.
20. *Nacorusana nigrescens* Miller 1954. 2·00 mm.
21. *Margasus afzelii* Stål 1855. 2·50 mm.

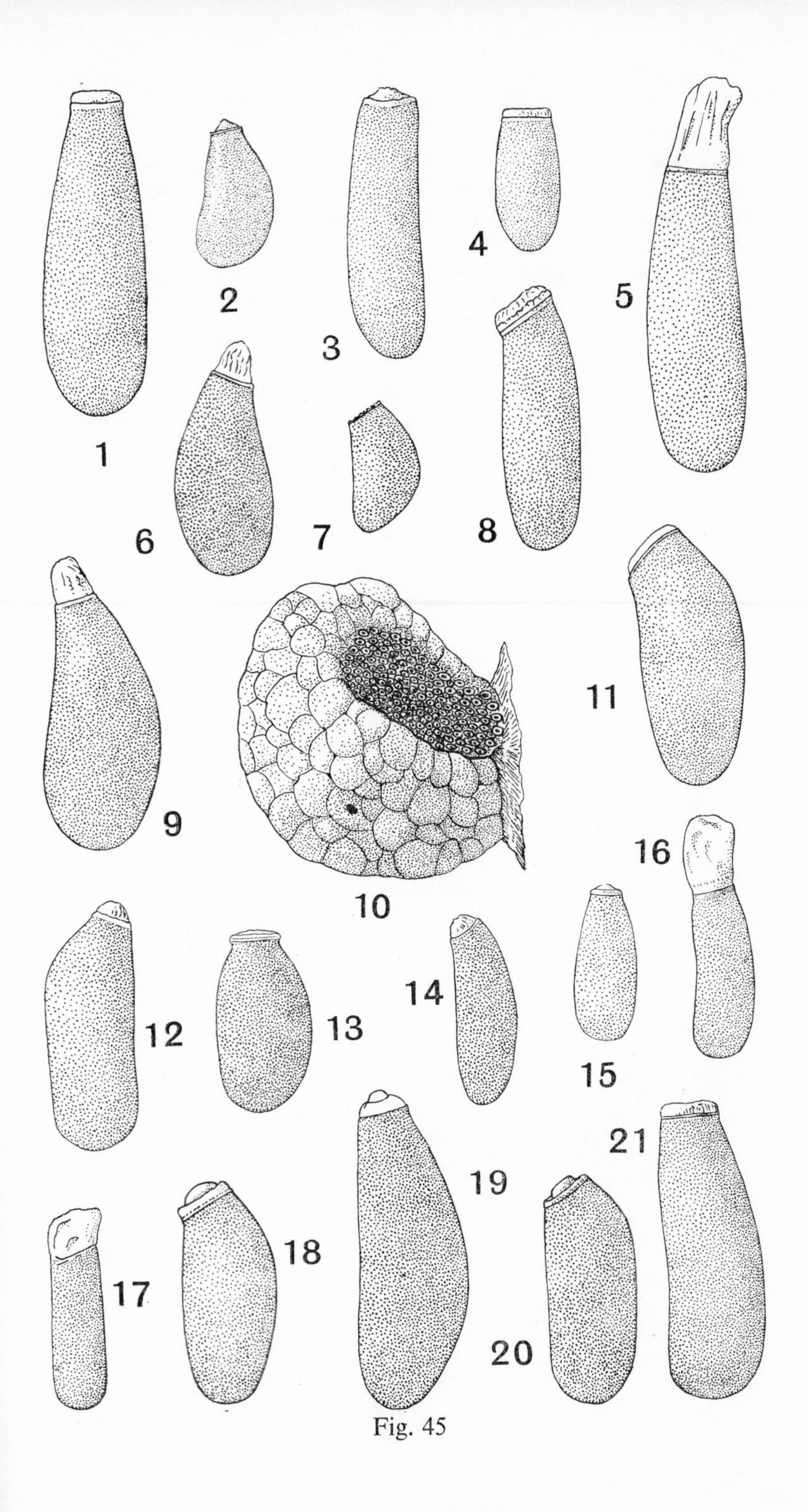

Fig. 45

1855, a brightly coloured species distributed throughout tropical Africa. I have taken this species at light. It is noteworthy that it has large and somewhat prominent ocelli reminiscent of those of some exclusively nocturnal genera belonging to other subfamilies.

As a rule **Harpactorinæ** are found mostly on low vegetation, but certain species, namely, *Pantoleistes princeps* Stål 1853 and *Nagusta subflava* Distant 1903 have been found on branches at some distance from the ground, the former at about twenty feet and the latter at eight feet. Another habitat recorded is that under loose bark of dead trees. Here, adults of *Havinthus pentatomus* Herrich-Schaeffer 1853 and neanides of *Sphagiastes* have been found. It is not improbable that *Henricohahnia* Breddin 1900, *Nyllius* Stål 1859, *Orgetorixa* China 1925 and allied genera, at least in the neanidal stages, live in such a habitat. The genus *Piestolestes*, to

---

Fig. 46 (*facing*)

Ova of Reduviidæ, Nabidæ, Miridæ, Tingidæ, Schizopteridæ, Microphysidæ, Belostomatidæ, Corixidæ, Nepidæ, Hydrometridæ, Gerridæ, Pleidæ, Gelastocoridæ, Mesoveliidæ, Saldidæ, Notonectidæ

1. *Gminatus wallengreni* Stål 1859. 1·50 mm. Reduviidæ-Harpactorinæ.
2. *Hiranetis* sp. 1·80 mm. Reduviidæ-Harpactorinæ.
3. *Vadimon bergrothi* Montandon 1892. 1·60 mm. Reduviidæ-Harpactorinæ.
4. *Sclomina erinacea* Stål 1861. 2·00 mm. Reduviidæ-Harpactorinæ.
5. *Nagusta* sp. 1·50 mm. Reduviidæ-Harpactorinæ.
6. *Campsolomus strumulosus* Stål 1870. 1·30 mm. Reduviidæ-Harpactorinæ.
7. *Pirnonota convexicollis* Stål 1859. 1·30 mm. Reduviidæ-Harpactorinæ.
8. *Arbela* sp. 1·70 mm. Nabidæ-Nabinæ.
9. *Gorpis papuanus* Harris 1939. 3·00 mm. Nabidæ-Gorpinæ.
10. *Psilistus chinai* Harris 1937. 2·40 mm. Nabidæ-Prostemminæ.
11. *Nabicula subcoleoptratus* Kirby 1837. 2·00 mm. Nabidæ-Nabinæ.
12. *Strongylocoris leucocephalus* Linnæus 1761. 2·50 mm. Miridæ-Orthotylinæ.
13. *Calocoris* sp. Miridæ-Mirinæ.
14. *Dictyonota stichnocera* Fieber 1844. 2·00 mm. Tingidæ-Tinginæ.
15. *Vilhennanus angolensis* Wygodzinsky 1950. 0·75 mm. Schizopteridæ (after Wygodzinsky).
16. *Myrmedobia tenella* Zetterstedt 1840. 0·47 mm. Microphysidæ (after Carayon).
17. *Lethocerus indicum* (Lepeletier and Servile) 1825. 3·70 mm. Belostomatidæ.
18. *Agraptocorixa eurynome* (Kirkaldy) 1922. Corixidæ (after Hale).
19. *Cymatia americana* Hussey 1922. Corixidæ.
20. *Sigara* (*Vermicorixa*) *alternata* (Say) 1825. Corixidæ (after Hungerford).
21. *Trichocorixa verticalis* (Fieber) 1851. Corixidæ.
22. *Nepa apiculata* Uhler 1862. Nepidæ.
23. *Ranatra fusca* Palisot Beauvois 1805. Nepidæ.
24. *Hydrometra murtini* Kirkaldy 1900. 1·00 mm. Hydrometridæ.
25. *Gerris* sp. Gerridæ.
26. *Plea striola* Fieber 1844. Pleidæ.
27. *Gelastocoris oculatus* (Fabricius) 1798. Gelastocoridæ.
28. *Mesovelia mulsanti* White 1879. Mesoveliidæ.
29. *Salda anthracina* Uhler 1877. Saldidæ.
30. *Notonecta irrorata* Uhler 1878. Notonectidæ.

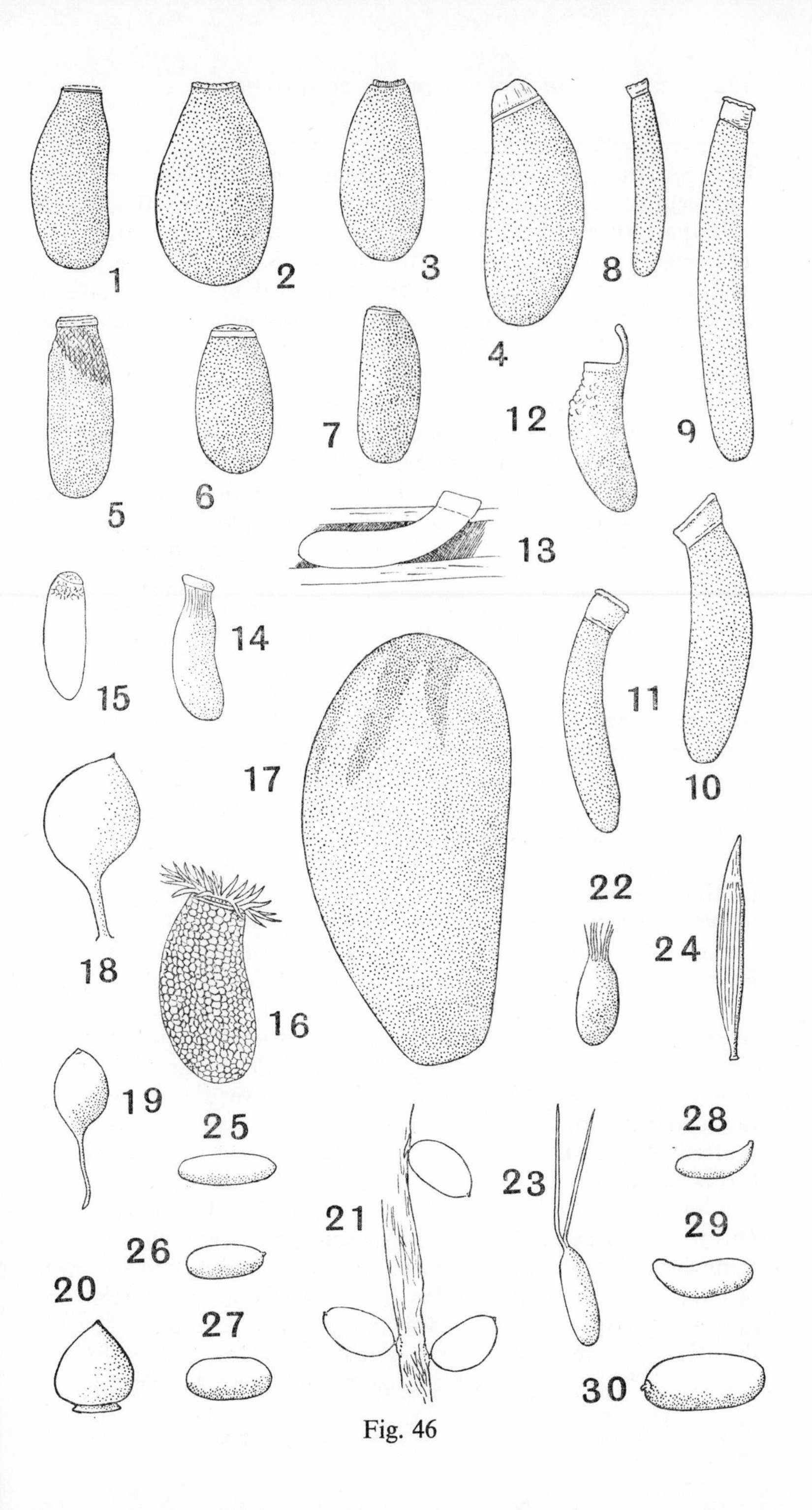
1
2
3
4
5
6
7
8
9
10
11
12
13
14
15
16
17
18
19
20
21
22
23
24
25
26
27
28
29
30

Fig. 46

judge by its flattened habitus, is probably another subcorticolous Reduviid. In the female the genital segments are somewhat produced and seem to be adapted for the insertion of ova into soft material.

Harpactorinæ are active insects and fly readily when disturbed. The flight is sustained for a short while only. Some fall to the ground from their resting place on the approach of danger. Some take to flight and travel a fairly long distance before alighting, others, such as *Phonoctonus nigrofasciatus*, often fly upwards to a considerable height, then circle round to alight more or less on the spot from which they departed.

The food of **Harpactorinæ** consists of other insects and their larvæ. There are recorded instances of adults devouring their own ova. I have noticed more specialized feeding in Southern Rhodesia in the case of *Rhinocoris albopunctatus*, *R. segmentarius* and *R. neavei* Bergroth 1912, the prey being honey-bees. Genera which have the second and third segments of the rostrum straight, such as *Henricohahnia*, *Orgetorixa*, *Nyllius* and allies and *Vadimon* Stål 1865 may possibly prey on larvæ living in burrows, the shape of the rostrum lending itself to probing. Prey is captured by seizing it with the anterior legs and piercing it with the stylets. When the victim has succumbed to the effects of the saliva and is inert, the bug continues to carry it attached to the rostrum until it has completely absorbed the body-fluids, whereupon the prey is released. The saliva is virulent and has been noted as being particularly so in *Coranus carbonarius* and species of *Rhinocoris*.

With regard to oviposition, a large number, probably the majority of **Harpactorinæ**, deposit their ova in groups containing variable numbers. From three or four to a hundred or so. Some species arrange their ova in single rows (*Coranopsis vittata* Horvath 1892), in parallel rows (*Vitumnus scenicus* Stål 1865) or singly (*Peprius pictus* Miller 1950). *Phonoctonus nigrofasciatus* arranges its ova in rings around a plant stem. All these species are Ethiopian. In the case of those species that deposit their ova in groups, the female secretes a glutinous substance and spreads it over the ova forming the periphery of the mass. This substance may afford some protection against parasitization by Hymenoptera or may prevent loss of moisture.

A special kind of method of arrangement of ova is practised by *Isyndus hero* (Fabricius) 1803, a Malayan species (Fig. 45, 10). This Reduviid, when about to oviposit, secretes a substance in the form of bubbles which rapidly harden and form together a sub-structure into which the ova are extruded and grouped close together with the longer axis vertical. The ova are entirely enclosed except for the opercula. This type of egg-mass might, with some justification, be termed an ootheca. It might be imagined that when the ova are covered almost entirely in this manner they would be protected from the depredations of hymenopterous parasites, but this is not always the case. Parasites have been obtained from such egg-masses and it is a curious fact that parasites, when, on becoming adult, wish to emerge from the mass, they do not select the shortest and easiest route, that is upwards through the ovum, but gnaw a passage through the sub-structure. The ova and methods of oviposition of other species of *Isyndus* have not yet been described.

The ova of **Harpactorinæ** do not exhibit a great variety of form, most of them being cylindrical, straight or feebly curved, or sub-ampulliform to a varying degree. The differentiated portion surrounding the operculum is sometimes relatively long, notably in the ovum of some *Rhinocoris* species. Some of the genera producing ampulliform ova and placing them singly or in small groups, attach the ovum by one side so that the longer axis is parallel more or less to the substratum. Certain ova, particularly those deposited in compact groups, have a very complicated operculum. A good example of this is the operculum of *Rhinocoris neavei*. This is a hollow cylinder, constricted near the base and apex with a small reticulate area basally. The apical margin is recurved, fimbriate or dentate. Within the cylinder is the opercular process, bluntly conical with filaments apically and with the surface minutely punctate and reticulate. In ova having this type of operculum the differentiated portion of the chorion is reflexed apically and covers the apex of the cylinder. The object of this type of operculum is to allow air to penetrate the ovum. Ova are also deposited in groups by *Panthous dædalus* Stål 1863, and they are covered by the female with a glutinous substance.

Neanides hatching from ova deposited in groups are usually gregarious until after the first ecdysis when they disperse and begin to feed. Since the duration of the first instar is short, abstention from food is apparently unimportant. It is possible that moisture in the form of dew or raindrops is imbibed occasionally.

Apiomerinæ. A subfamily containing many genera, all of them distributed in the Nearctic and Neotropical Regions. Most of them are dull in colour, but some species, namely, *Apiomerus binotatus* Champion 1899, *A. elatus* Stål 1862, *A. fasciatus* Herrich-Schaeffer 1848, *A. elegans* Distant 1903 and *Ponerobia bipustulata* (Fabricius) 1781 have red and yellow markings. *Apiomerus vexillarius* Champion 1899, *A. geniculatus* Erichson 1848, *A. ochropterus* Stål 1866 and *A. pilipes* (Fabricius) 1787 have two red foliaceous expansions arising from the eighth abdominal segment. In *A. nigrolobus* Stål 1872 these expansions are black.

The **Apiomerinæ** are characterized by long anterior tibiæ which, like the median and posterior pair are strongly setose. The anteocular region of the head is not much shorter than the postocular which is somewhat widened in the region of the ocelli which are lateral in position. A stridulatory furrow is present in all genera. The anterior and median tarsi are very short and when not in use rest in a sulcus on the upper surface of the tibia, the sulcus varying in depth; for example, in *Ponerobia* Amyot and Serville 1843 it is very shallow. In *Micrauchenius* Amyot and Serville 1843 there is no sulcus. In most genera the median tarsi are normal but, in *Heniartes* Spinola 1837 they are very short. In *Amauroclopius* Stål 1868 the anterior tarsi are extremely small.

The habit of covering the anterior tibiæ with a resin for the purpose of capturing prey is characteristic of some members of this subfamily, but how many practise this method is not known. Specimens in collections are often found to have a considerable amount of vegetable and other debris adhering to them on account of a glutinous material on their legs and body. Whether this substance is secreted by the insect or is deliberately collected by it is not always clear. The actual collection of a resinous substance by a member of this subfamily was actually observed in the case of *Beharus lunatus* Lepeletier and Serville 1825 in Surinam by Uittenboogaart.

Some genera, among which *Heniartes*, have small patches of a white wax-like substance on the body similar to those to be seen in many **Harpactorinæ.** It is not known how or from where this is produced. Possibly it may have covered a much greater area of the insect prior to the final ecdysis and may, indeed, have played a rôle in that operation.

The **Apiomerinæ** are diurnal and frequent foliage and flowers in forest clearings.

The ova and developmental stages of some species of *Apiomerus* Hahn 1831 and *Heniartes* have been studied. *Apiomerus spissipes* (Say) 1825 and *Heniartes jaakoi* Wygodzinsky 1947 deposit their ova in a mass, the former covering the ova forming the outer row of the mass with moderate amount of glutinous substance. The latter embeds them in a copious quantity of the same kind of substance so that the shape is entirely concealed. The substance is secreted in a spongy state but it rapidly becomes hard. The ova of known species of **Apiomerinæ** are cylindrical with a differentiated portion similar to that of **Harpactorinæ.**

Diaspidiinæ. This subfamily contains three genera, *Rhodainiella* Schouteden 1913, *Diaspidius* Westwood 1857 and *Cleontes* Stål 1874, all distributed in the Ethiopian Region.

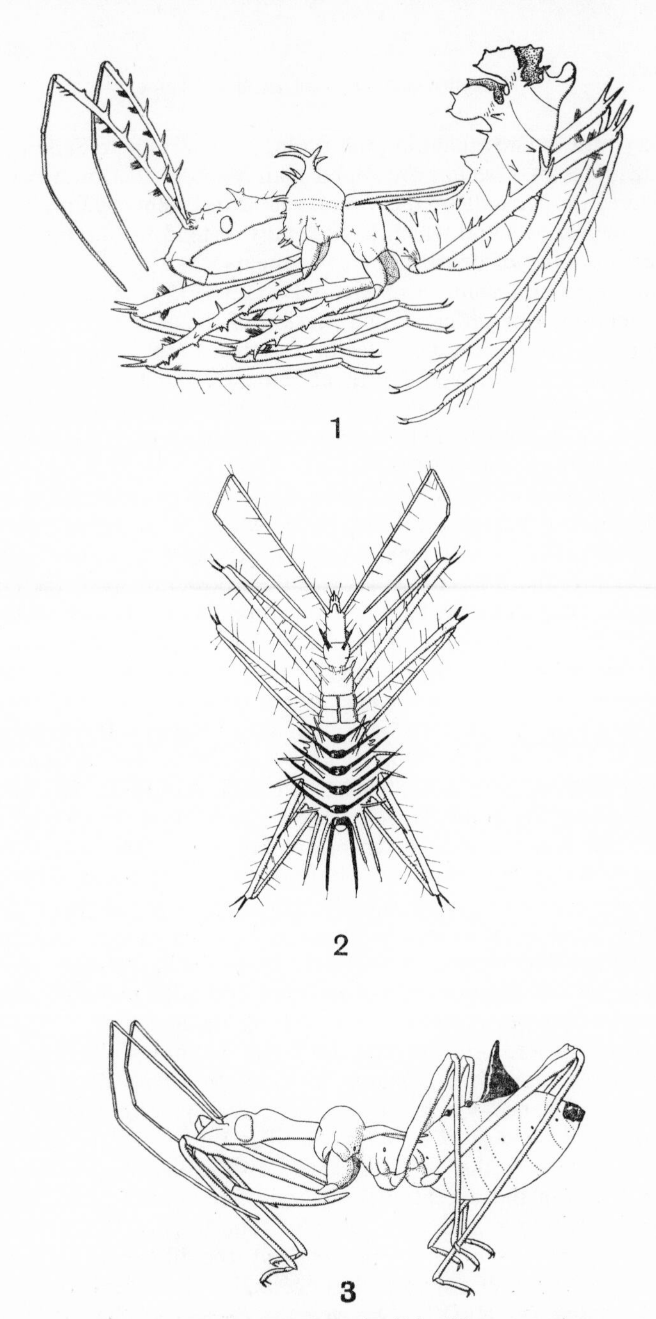

Fig. 47. Neanides of Reduviidæ-Harpactorinæ

1. *Sphagiastes ramentaceus* (Germar) 1837. 5th instar.
2. *Hoffmannocoris chinai* Miller 1950. 1st instar.
3. *Pantoleistes* sp. 4th instar.

Very little is known about their ecology. According to Schouteden the neanides of *Cleontes* are found mainly under the loose bark of trees and on account of the viscous substance they accumulate on their anterior legs and which, incidentally, extends to the body also, a great quantity of debris adheres to them.

The ovum of *Cleontes ugandensis* Distant 1912 is cylindrical with a moderately long differentiated portion of the chorion.

Dorsal glands in the neanides of *Diaspidius* and *Rhodainiella* are located on the fourth and fifth abdominal segments; in *Cleontes* a single gland ostiole is present on the third segment (Fig. 26, 4). In the adult of this genus the position on the dorsum of the abdomen is indicated by a sub-erect tubercle.

The diagnosis of the subfamily is as follows:—macropterous; head and legs setose; rostrum thick, straight, extending to anterior margin of prosternum; basal segment very short; ocelli lateral, elevated; anterior pronotal lobe much shorter than posterior lobe, the latter strongly produced posteriorly and concealing the scutellum; connexivum expanded, simple or undulate; discal cell of corium large; legs moderately thick; anterior tarsi reduced or absent; when present rest in a sulcus at the apex of the tibia.

Ectinoderinæ (Plate IV). This subfamily contains three genera: *Ectinoderus* Westwood 1843, *Amulius* Stål 1865 and *Parapanthous* Distant 1919, which are confined to the Indo-Australian Region.

They have the following characters—anteocular portion of the head very short; postocular considerably longer and transversely enlarged in ocellar area; pronotum strongly flattened with the posterior lobe much longer than the anterior lobe and with a median excision apically, thus not concealing the apex of the scutellum; connexivum not expanded; anterior tibiæ thick, strongly setose; anterior tarsi composed of one segment and with modified claws; hemelytra fully developed and with a large discal cell.

Ectinoderus and *Amulius* and, so far as is known, *Parapanthous* obtain their prey with the anterior tibiæ to which they apply a resinous substance yielded by certain trees among which *Pinus merkusii* and *Agathis alba*. With the tibiæ thus covered they lie in wait, usually with the body directed downwards on a tree trunk. More details of this exceptional method of capture of prey by these Reduviids will be found in the chapter 'The Legs of Heteroptera'.

With regard to the genus *Parapanthous*, the difference between it and *Amulius* is minor only, which suggests that they are synonymous.

Phonolibinæ (Plate IV). Two genera, *Phonolibes* Stål 1854 and *Lophocephala* Laporte 1832, are contained in this subfamily; the former is distributed in the Ethiopian Region and the latter in India and Ceylon. *Phonolibes* has a somewhat elongate head, constricted immediately behind the eyes, widely separated ocelli, straight rostrum with three visible segments, the articulation between the second and third segments being very feebly demarcated, a fact that has lead to the opinion that the rostrum is composed of two segments only: stridulatory furrow and harpagones absent; legs and body usually with abundant, short secretory hairs. *Lophocephala* has similar characters but a stridulatory furrow is present. Alary polymorphism occurs in both sexes. Dorsal glands are present in the neanides on the third, fourth and fifth segments of the abdomen.

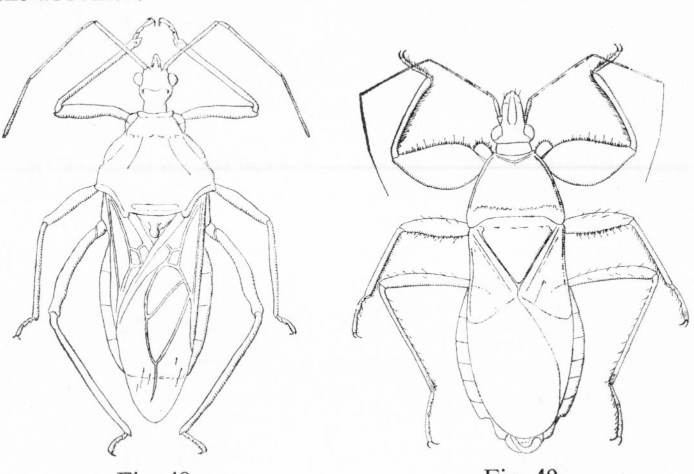

Fig. 48
Tegea pinguescens Miller 1941 (Reduviidæ-Tegeinæ).

Fig. 49
Pachynomus biguttatus Stål 1863 (Pachynomidæ).

Phonolibes are nocturnal and remain concealed in the daytime under logs lying on the ground. They have also been found in the nests of termites.

The ovum of *Phonolibes obsoletus* Horvath 1914, which is deposited singly by the female, is ampulliform with a minutely granulose chorion. Nothing is known of the habits or developmental stages of *Lophocephala*.

Tegeinæ (Fig. 48). This subfamily contains four genera; *Tegea* Stål 1863, *Tegellula* Breddin 1912, *Nannotegea* Miller 1953 and *Campylorhyncha* Stål 1870, all confined to the Indo-Australian Region. The species belonging to these genera are characterized by having the anterior lobe of the head not much longer than the posterior lobe, the eyes relatively large, the ocelli small and widely separated, rostrum straight and slender, composed of three segments, the division between segments two and three not well-defined;

integument usually glabrous; pronotum with two sinuate dorso-lateral carinæ; posterior lobe much longer than anterior lobe, scutellum with a thick apical spine. The rostrum of *Campylorhyncha* is somewhat different, being relatively thick and the apical segment is somewhat recurved outwards.

The prosternal furrow is somewhat obscurely striate, particularly at the middle where also the striæ are wider than those at the posterior end. In *Tegea atropicta* Stål 1863, the striæ which are particularly obscure are confined to part of the basal third of the furrow and are absent from the rest of the furrow. The rostrum in this species extends almost to the posterior margin of the mesosternum where there are somewhat indefinite transverse sulci, between which the ridges are rounded.

The habitats of **Tegeinæ** are the trunks of trees in forest areas. The ova and methods of oviposition are unknown. There is no information regarding their food but, from the shape of the rostrum it seems likely that they search for their prey by probing crevices and insect-borings in the bark of trees.

References

Abalos and Wygodzinsky 1951; Blanchard 1902; Brindley 1930; Brumpt 1912, 1914a, 1914b; Buxton 1930; Carayon, Usinger and Wygodzinsky 1958; China 1926; China and Usinger 1949; Costa Lima, Campos Seabra and Hathaway 1961; Dispons 1951, 1955a, 1955b; Edwards 1962; Gillett 1957; Handlirsch 1897; Hase 1933, 1941; Herrer, Lent and Wygodzinsky 1954; Jacobson 1911; Kershaw 1909; Kirkaldy 1911; Miller 1952, 1953b, 1954a, 1954b, 1955a, 1955b, 1955c, 1956b, 1956c, 1956d, 1957a, 1958b, 1959; Odhiambo 1958; Readio 1926, 1927, 1931; Roepke 1932; Ryckman 1954, 1961, 1962; Schouteden 1931; Shun-ichi Nakao 1954; Southwood 1955; Usinger 1943, 1944, 1946, 1958; Usinger and Wygodzinsky 1964; Verhoeff 1893; Villiers 1945, 1949; Wygodzinsky 1946, 1947, 1948; Wygodzinsky and Usinger 1963.

PACHYNOMIDÆ (Stål) 1873, *Enum. Hemipt.* **3**, 107 (Fig. 49)

Formerly included in the Nabidæ but subsequently considered to belong to the Reduviidæ as a subfamily, since they have the same type of male genitalia and odoriferous glands; the rostrum is composed of four visible segments; stridulatory furrow lacking; ocelli present but reduced in some genera; antennæ composed of five segments. A small *fossula spongiosa* is present on the anterior and median tibiæ.

In *Pachynomus* Klug 1830, *Camarochilus* Harris 1830 and *Punctius* Stål 1873 the ocelli are greatly reduced; in *Aphelonotus* Uhler 1894 they are relatively large.

Pachynomus has strongly incrassate anterior femora, the lower surface of which is armed with abundant, short, slender denticles with rounded apices. The anterior and median tibiæ and the median femora are similarly armed, the denticles on the tibiæ being short and rounded.

The Pachynomidæ are presumably mainly nocturnal, and during the daytime seek seclusion under stones and bark.

They are distributed in parts of Europe, Asia and Africa.

VELOCIPEDIDÆ Bergroth 1891, *Wien. ent. Zeit.* **10,** 265 (Fig. 50)

This family contains only one genus of which four species are known at present. They are medium sized, broadly oval insects with a long rostrum composed of three segments and with slender antennæ composed of four segments. They have prominent eyes, a wide scutellum with a distinct cuneus and wide embolium.

The females have an ovipositor and, although nothing is known of their habits and life history, it may be assumed that they insert their ova into plant tissues. Possibly not more than eight or ten ova develop at one time. The ovum of *Scotomedes alienus* (Distant) 1904 is relatively large, cylindrical, narrow, feebly curved and with a narrow differentiated portion. This species is distributed in Sikhim, Burma and Indo-China, but others yet undetermined, have been collected in Sarawak, Dutch New Guinea and in the Philippine Islands.

The relationship of the family has been discussed by several authors including Stål, Bergroth, Distant, Kirkaldy, Reuter and Blöte; the last-mentioned considered that it is allied to the Nabidæ. This view is probably correct.

References

Blote 1945; Distant 1904.

MEDOCOSTIDÆ Stys 1967, *Acta ent. bohemoslov* **64,** 439–465 (Fig. 51)

This family includes two species, *Medocostes lestoni* Stys and *M. carayoni* Stys, the former from Ghana and the latter from the Congo. Each species measures less than 10 mm. in length.

According to Stys the 'Medocostidæ differ from the Cimicimorpha by segmental ratios of the labium and by construction of the buccular region, and together with the Velocipedidæ also, by the presence of a free epipharangeal projection. They represent probably a sister-group of Velocipedidæ, and, as indicated by similar structure of the inner female genitalia and venation of the hemelytra, they are apparently closely allied to the ancestors of the Nabidæ.'

A general diagnosis of the family shows the representatives to be strongly sclerotized, with an elliptical body, having the dorsal surface almost flat and the ventral surface convex. Head, pronotum, mesoscutellum, pleura (partly), clavus and corium, ventral surface of the abdomen and most of the dorsal surface of the abdomen densely alveolate. The head is almost porrect, the eyes large, sessile, the ocelli well-developed. The antennæ have four segments. The pronotum is trapezoidal with a narrow collar. All acetabula open posteriorly. The lateral parts of the abdomen are vertically raised

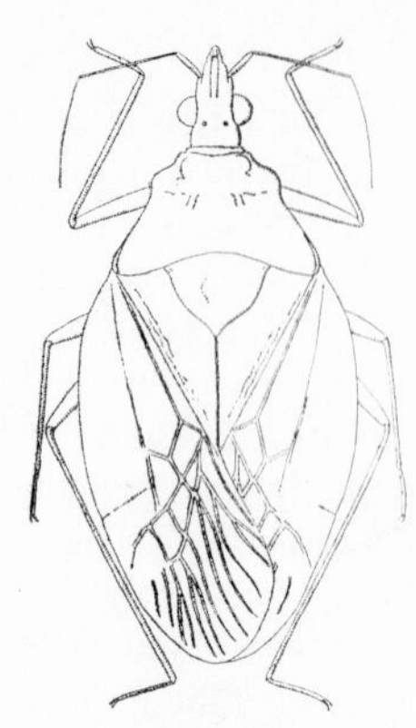

Fig. 50
Scotomedes alienus (Distant) 1904
(Velocipedidæ).

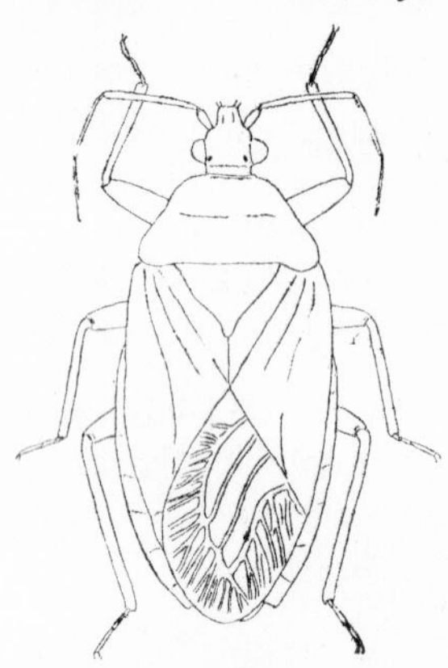

Fig. 51
Medocostes lestoni Stys 1967
(Medocostidæ) (after Stys).

much above the level of the hemelytra. Femora simple with the apical area of the ventral surface sulcate. Tibiæ slender. Tarsi with three segments. Corium with simple R and M running closely together, separated by a medial fracture only and with a simple Cu; membrane with three long basal cells and many veins radiating from them.

Both species of *Medocostes* were found in localities situated in transition areas between the tropical rain-forest and tropical bushy grassland savannah zone. They have been captured in light traps. Nothing is known, so far, of their habits and food-preferences.

Reference

Stys 1967.

NABIDÆ Costa 1852, *Cimic. Regni Neap. Cent.* **3**, 66 (Figs. 45, 52)

There are five subfamilies of Nabidæ, namely, **Nabinæ** Reuter 1890 with a broad pronotal collar, sub-membranous hemelytra with the clavus widened posteriorly, the commissure being longer than the scutellum, and the anterior coxæ open posteriorly. This is a large subfamily, members of which inhabit low vegetation, grasses and bushes. Distribution, world-wide. **Prostemminæ** Reuter 1890 with the anterior coxæ open posteriorly, the pronotal collar reduced, the clavus narrowed posteriorly with the commissure shorter than the scutellum. A small subfamily of brightly coloured, terrestrial species. Distribution, world-wide. **Gorpinæ** Reuter 1909, with the pronotum not strongly convex, the abdomen wide basally, the median femora in the male not incrassate, the anterior coxal cavities closed posteriorly and the anterior coxæ very long. Distribution, North America. **Arachnocorinæ** Reuter 1890 with the pronotum very strongly convex, laminately produced particularly on each side of the scutellum, the median femora in male incrassate and spined and the abdomen narrowed basally. Distribution, tropical America. **Carthasinæ** Blatchley 1926. In this subfamily, representatives have the tibiæ with a spatulate process apically below insertion of tarsi, the lower surface of the head with two pairs of spines and the lower surface of the anterior femora with spines and the usual bristly pubescence. Ocelli lacking. Distribution, tropical America.

This family is composed of mainly small or very small dull-coloured insects with predaceous habits. There are, however, certain genera which have some part of the body brightly coloured, the colours mainly red and yellow. Among those so coloured may be mentioned, *Prostemma* Laporte 1832 (**Prostemminæ**), *Aristonabis* Reuter and *Poppius* 1909, *Alloeorhynchus* Fieber 1861 (**Nabinæ**).

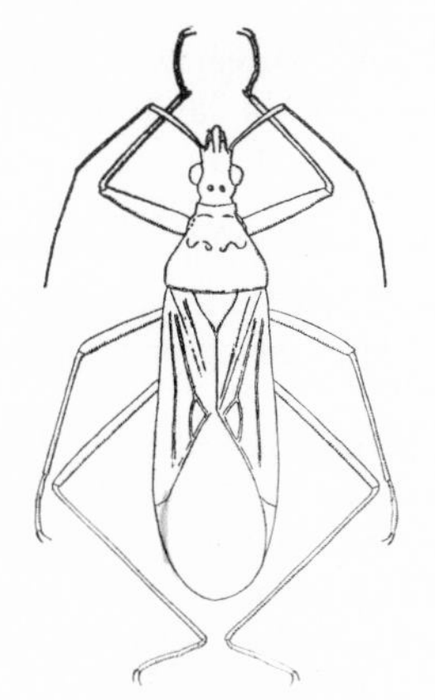

Fig. 52
Nabis ferus Linnæus 1761
(Nabidæ-Nabinæ).

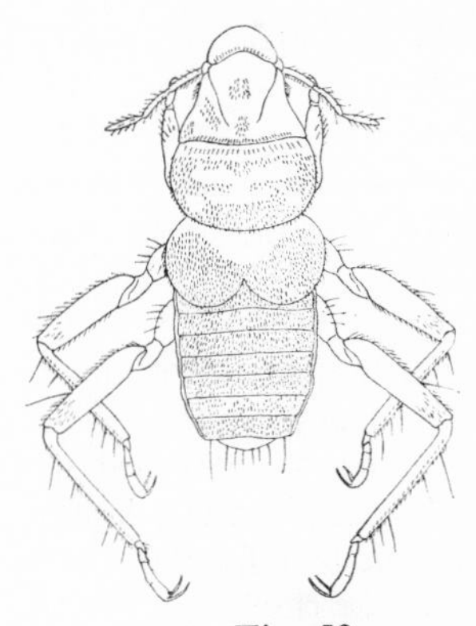

Fig. 53
Hesperoctenes impressus Horvath 1910
(Polyctenidæ).

The Nabidæ are closely related to the Reduviidæ and were formerly placed in that family as a subfamily. They may be distinguished from the great majority of Reduviidæ mainly by the rostrum which has four visible segments. Most of the Nabidæ are fully alate, but brachypterous forms occur in several genera, for example, in *Nabicula* Kirby 1837, *Dolichonabis* Reuter 1908, *Reduviolus* Kirby 1837, *Hoplistoscelis* Reuter 1890 *Vernonia* Buchanan White 1878, *Alloeorhynchus* (**Nabinæ**) and *Prostemma.*

The legs are mostly slender and setose. In *Gorpis* Stål 1859 the anterior legs are similar to an Emesine type and the males of some species of *Arbela* Stål 1865 (**Nabinæ**) have a dense tuft of setæ near the base of the posterior tibiæ.

Nabidæ are found mostly on plants and some of them insert their ova into plant tissues. The ova of several species have been described (Fig. 46). They are mostly cylindrical, somewhat slender, straight or feebly curved, with a short differentiated portion of the chorion.

Alloeorhynchus and *Phorticus* Stål 1858 live mainly in the soil among plants and vegetable debris, the latter being occasionally found on shrubs. Many Nabidæ are nocturnal and are often attracted to artificial light. *Arachnocoris albomaculatus* Scott 1881 (**Arachnocorinæ**) has been recorded in association with spiders.

References

Carayon 1950b, 1954; Myers 1925; Reuter and Poppius 1909; Southwood 1953.

POLYCTENIDÆ Westwood 1874, *Thesaur. Ent.* 197 (Fig. 53)

This family is allied to the Cimicidæ and parasitic on bats. In general habitus the Polyctenidæ are somewhat similar to dipterous bat-parasites of the family Nycteribiidæ; in fact, the first known Polyctenid *Polyctenes molossus* Giglioli 1864 was placed in that family. Some years later, Westwood proposed the family Polyctenidæ and placed it in the Order Anopleura. In 1879 Waterhouse placed the family again in the Diptera but, a year later, he agreed (although with some doubt) to its hemipterous affinities.

The fact that the Polyctenidæ should be considered hemipterous was established by Speiser, who placed it near the Cimicidæ, where it now stands. With regard to reproduction, in view of the close relationship between this family and the Cimicidæ, it would not be unreasonable to infer that the methods of copulation were similar. This is not the case, however.

In the Cimicidæ it has been demonstrated that the spermatozoa are received through a slit or emargination of the posterior margin of the fourth or fifth visible segment. The spermatozoa gradually reach the hæmocoel, whence they make their way to the oviduct and the ova.

Jordan has pointed out that the intromittent organ of the males of Cimicidæ and Polyctenidæ are similar, which might suggest the presence in the Polyctenidæ of an 'organ of Berlese', but he shows that the structure of the female abdomen does not indicate that copulation is effected in the manner common to the Heteroptera. More observations are essential before the problem of the method of fertilization is resolved and, in view of the habits of these insects a solution is not likely to be easily obtained. It is noteworthy that in the Polyctenidæ the early neanidal stages are passed in the body of the female.

Polyctenidæ are somewhat rare in collections, thus the probable distribution cannot be stated precisely. There is little doubt, however, that the family occurs in tropical and sub-tropical regions.

References

Ferris and Usinger 1939; Hagen 1931; Jordan 1911, 1913; Speiser 1904; Waterhouse 1879.

CIMICIDÆ (Latreille) 1804, *Hist. Nat. Crust. Ins.* **12**, 235 (Fig. 54)

Parasitic insects with a compressed habitus and without metathoracic wings; ocelli absent; clypeus triangular broadening apically to truncate apical margin, except in *Primicimex*, in which case the rostrum very short and not reaching level of insertion of antennæ; hemelytra always rudimentary; male genitalia asymmetrical; female with opening to Ribaga's organ on ventral or dorsal surface of abdomen; eyes small; abdomen oval in outline.

Copulation in members of this family is of a specialized type. It is effected by the male introducing the penis into a special pouch-like organ—Ribaga's organ—which has its orifice on the dorsal and ventral surface of the fourth abdominal segment, the position being of a generic character. Part of the spermatozoa penetrates the walls of this pouch and, by means of the hæmocoel reaches the *receptaculum seminis* of the female, while another part penetrates directly into the ovarioles. Yet another part is digested by the amœbal cells which group around the organ and the products of the digestion contribute eventually to the development of the ovaries which are not completely mature at the time of copulation.

The Cimicidæ are photophobic. In this family is the well-known bed-bug, the widely distributed human ectoparasite, *Cimex lectularius* Linnæus 1758, and many related species which are ectoparasites of birds and bats. *C. lectularius* and other members of the family conceal themselves as much as possible in cracks and crevices in woodwork of houses where they deposit their ova.

The Palæarctic Reduviid *Reduvius personatus* Linnæus 1758 is said to prey on the bed-bug, and in the Malaysian sub-Region there is evidence which, however, has not been confirmed definitely that it is attacked by another Reduviid, *Vesbius purpureus* Thunberg 1784, a small, brightly coloured species often found under the floor boards of houses of the Malay type; that is, raised on poles and about six or seven feet from the soil. This Reduviid has been seen in a concentration camp in Sumatra by the writer, where bed-bugs were present in thousands.

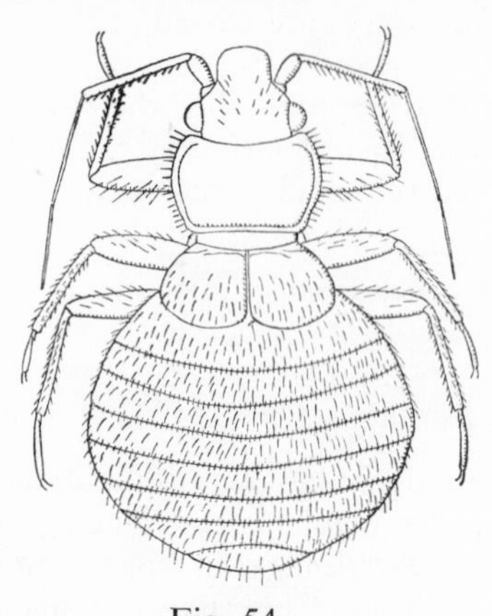

Fig. 54
Loxaspis mirandus Rothschild 1912 (Cimicidæ).

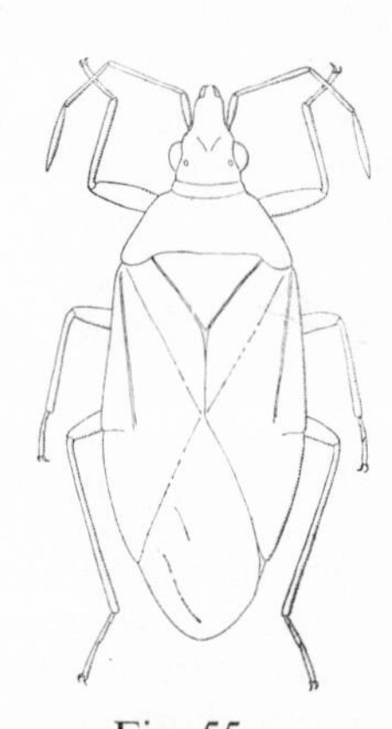

Fig. 55
Acompsocoris pygmæus Fallèn 1807 (Anthocoridæ).

An egg-burster is present in the ova of Cimicidæ. It consists of a V-shaped series of spines on the embryonic cuticle. A similar arrangement of spines has been observed in the egg-burster of certain Reduviidæ. The ovum of *C. lectularius* is white, cylindrical, narrowed at each end and is deposited singly or in small groups.

C. lectularius, *C. hemipterus* Fabricius 1803, *Oeciacus hirundinis* (Jenyns) 1839 (**Cimicinæ**) have been recorded as vectors of *Trypanosoma cruzi*, the causative agent of Chaga's Disease in South America.

The subfamilies of the Cimicidæ are: **Hæmatosiphoninæ** Jordan and Rothschild 1912, in which the bristles on the lateral margins of the pronotum are not serrate at their outer edges and are usually cleft or dentate apically, the tibiæ have short bristles as well as short hairs and the organ of Ribaga is usually dorsal. Distribution U.S.A. **Primicimicinæ** Usinger 1966 have the clypeus narrowed apically, the rostrum very short not reaching the level of insertion of the antennæ. Distribution, North America. **Cimicinæ** Van Duzee 1916, in which the bristles on the pronotum laterally are serrate on their outer edges, the metasternum widened posteriorly, plate-like between posterior coxæ and the organ of Ribaga is situated ventrally on the fourth and fifth segments of the abdomen. Distribution,

Cosmopolitan. **Cacodminæ** Kirkaldy 1899, in which the tibiæ have fine setæ which may be short or very long but not stiff and spine-like. Distribution, Ethiopian and Oriental Regions.

The **Hæmatosiphoninæ** are parasites of birds, including the domestic fowl, the **Primicimicinæ** have been found on cave bats, the **Cacodminæ** also have been found on bats, while the **Cimicinæ** are parasites of birds, bats and man.

References

Bacot 1921; Carayon 1953; Hase 1917; Jordan 1922; Sikes and Wigglesworth 1931; Ryckman 1958; Southall 1730; Usinger 1947b, 1959 1966; Wendt 1939.

ANTHOCORIDÆ (Amyot and Serville) 1843, *Hist Nat. Hémipt.* XXXVII, 262 (Fig. 55)

The Anthocoridæ are very small, mainly predaceous insects, some of which are cosmopolitan in range on account of their being transported from region to region in stored products such as rice and copra, among which they find their food which consists of the larvæ of coleopterous and other insect pests and of mites.

The characters of the Anthocoridæ may be summarized as follows: head produced, truncate and broadly rounded anteriorly, horizontal with the clypeus elongate, ocelli present in alate forms, antennæ with four segments, sometimes two basal segments clavate with the apical segments filiform or fusiform, pronotum usually trapeziform, convex anteriorly, depressed posteriorly, hemlytra in macropterous forms with an incomplete but distinct cuneus, corium and clavus not or only feebly punctate, membrane without a basal cell and with one to four veins, male genitalia asymmetrical, tarsi with three segments.

Certain species possess a special modification of the seventh ventral segment to which the name *omphalus* has been given. During copulation, the male organ is introduced into it and not into the genital passage of the female. The presence of this peculiar modification has been established in the genus *Cardiastethus* Fieber 1860, *Poronotellus* Kirkaldy 1904 and *Brachysteles* Mulsant and Rey 1852, but not in all species.

The conclusion has been reached that the omphalous character should be considered as practically deprived of systematic significance for taxonomic supra-specific categories. It appears to correspond to a particular stage of an evolutionary phenomenon, revealing itself in the same manner but with a different degree of rapidity in several neighbouring channels.

Both alate and apterous forms occur but in different species. The females are sometimes provided with an ovipositor. Those with ovipositors insert their ova into vegetable tissue. Ovoviviparity occurs in some genera. Up to the present time, not a great deal is known about their developmental stages or their habits, and information regarding their food, in certain instances, is based on flimsy evidence only.

The ovum of *Anthocoris confusus* Reuter 1882 (**Anthocorinæ**) according to Butler is short, cylindrical, rounded basally and apically and obliquely truncate apically. It is very pale yellow with a white collar. The ovum of *Anthocoris nemorum* Linnæus 1761 (**Anthocorinæ**) according to the same author is pale yellowish-white. It is cylindrical, rounded apically and basally and has a white collar of smaller diameter. The surface of the chorion is corrugated.

Physopleurella africana Carayon 1956 has been found in the nest of a weaver bird, *Ploceus cucullatus* in West Africa. Incidentally, in this genus a new instance of pseudoplacentary viviparity has been recorded. The genus *Cardiastethus* has been found under the bark of *Eucalyptus*, *C. inquilinus* China and Myers 1929 in the nest of an Oxyopid spider, *Xylocoris flavipes* (Reuter) 1875 in stored grains. This species and *Piezostethus flavipes* Reuter 1884 may be egg-predators.

Falda queenslandica Gross 1954 and several species of *Lasiochilus* Reuter 1871 frequent the debris on forest floors in Australia and *Scoloposcelis* spp. Fieber 1864 have been seen under the bark of leguminous trees, principally *Albizzia* in Africa.

Triphleps cocciphagus Hesse 1940 (**Anthocorinæ**) has been recorded as preying on the red scale of citrus, and *Montandoniola moraguesi* Puton 1896 (**Anthocorinæ**) is predaceous on the thrips *Gnaikothrips ficorum* Marchal, a leaf-roller of Ficus.

Other instances of the type of food of Anthocoridæ are aphids and the pine scale *Matsucoccus* eaten by *Anthocoris nemoralis* (Fabricius) 1794 (**Anthocorinæ**), the excrement of aphids by *Anthocoris gallarum-ulmi* de Geer 1773, ova of *Heliothis obsoleta* (Lepidoptera) devoured by *Triphleps insidiosus* Say 1831 (**Anthocorinæ**). The normal food of *Anthocoris nemorum* is apparently of an animal nature, but it has been accused of damaging leaves and shoots of the hop plant. At the same time, since it feeds on aphids which infest that plant, the damage it causes by feeding on the plant is somewhat counterbalanced. *Lyctocoris campestris* Fabricius 1794 (**Lyctocorinæ**) has been stated to be vegetarian in its diet, but later to have adopted a carnivorous diet and to feed on the blood of horses and cattle. It has been found in crevices in dwellings.

In Java *Triphleps* (*Orius*) *persequens* White 1877 (**Anthocorinæ**) and *Sesellius* (*Scoloposcelis*) *paralellus* Motschulsky 1863 (**Lyctocorinæ**) suck the young larvæ of *Scirpophaga* and *Proceras* (Lepidoptera).

Lasiochilus perminutus Poppius 1910 has been reported as an important enemy of the coleopterous borer of banana stems—*Cosmopolites* (Curculionidæ). *Piezostethus flavipes* Reuter 1875, a dark brown species measuring about 2 mm. in length, is repeatedly found in rice and tapioca stores in Java, Sumatra and Banka. There, it most probably preys on mites and Psocids. This species is often found in holds of ships and has also been detected in stored products in various parts of the world. Anthocoridæ are also found in vegetable debris, under bark and stones, in birds' nests and burrows of mammals.

There are three subfamilies: **Lyctocorinæ** Reuter 1884, in which the cells of the metathoracic wings may have a hamus which, if present, arises from the *vena connectens*; the apical antennal segments are usually filiform. Distribution, world-wide. **Anthocorinæ** Reuter 1884, with the cells of the metathoracic wings always provided with a hamus which arises from the *vena decurrens* or far from it and from the *vena subtensa*; the apical segments of the antennæ are fusiform and very rarely filiform. **Dufouriellinæ** Van Duzee 1916, in which the hamus is absent. Distribution, world-wide. Southwood and Leston consider the subfamily **Anthocorinæ** should be in the family Cimicidæ.

References

Butler 1923; Carayon 1952, 1953a, 1953b, 1956, 1957; Delamare Deboutteville and Paulian 1952; Gross 1955; Hesse 1940; Hill 1957; Leston 1954b; Poppius 1909; Reuter 1885; Sands 1957; Southwood and Leston 1959.

MICROPHYSIDÆ Dohrn 1859, *Cat. Hemipt.* 36 (Figs. 46, 56)

Members of this family are very small, dull-coloured insects, apterous or alate; tarsi with two segments; cuneal fracture present; eighth ventral segment in female divided into two plates on each side of ovipositor, when present; membrane with a single square cell at base between two longitudinal veins, rarely without such a cell or without venation; sexual dimorphism considerable.

The Microphysidæ are rare or uncommon and, on account of their small size, can easily be overlooked by collectors. Extremely little is known of their biology with the exception of a few details concerning their habitats. They are found mainly on old and dying trees covered with moss and lichens. The females of several species are also to be found under bark, in old brushwood, under moss and in ants' nests. The males may be found in the same sites or in vegetation in the vicinity of trees in which females are living.

Carayon, in an attempt to determine the food of *Myrmedobia tenella* (Zetterstedt) 1838 and of *M. coleoptrata* (Fallen) 1807, supplied females of both these species with Acarids, Collembola,

Psocoptera, Thysanoptera, Psyllidæ, larval microlepidoptera, but the attempt to gain information was in vain. On the contrary, females of *Loricula elegantula* (Baerensprung) 1858, several times captured and fed on small Psocoptera, especially *Reuterella helvimaculata* End., as well as *Lachesiella pedicularia* Linnæus. *Loricula* fed readily in captivity, but in the absence of prey were inclined to cannibalism.

The ovum of *Loricula elegantula*, according to Butler, is one-third of a millimetre long, broad, rounded posteriorly, truncate at right angles anteriorly, with dark brown crenulate free edge. Oviposition methods and a detailed description of the ovum of *Myrmedobia*

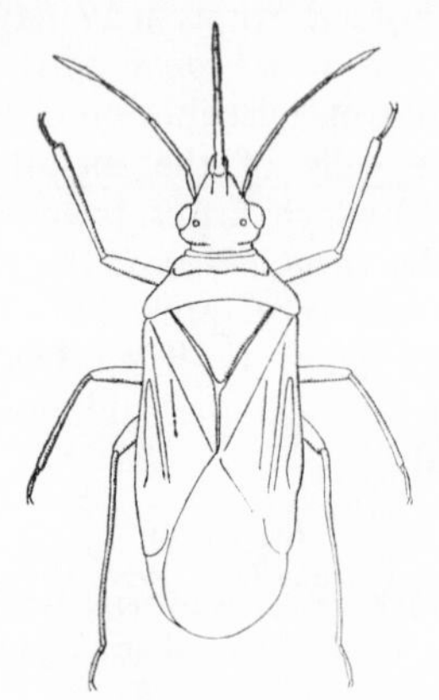

Fig. 56
Microphysa pselapiformis Curtis 1833
(Microphysidæ).

tenella have been recorded by Carayon (Fig. 46, 16). The ova are placed by the female among lichens or sometimes under the cracked and slightly raised bark of small branches. They are usually deposited singly or in groups of two. In shape they have the general form of the ova of the Cimicoidea, but are remarkable for the long chorionic processes which spread immediately after the ovum leaves the body of the female.

References

Butler 1923; Carayon 1949b; China 1953a.

PLOKIOPHILIDÆ (China), 1953, *Ann. Mag. Nat. Hist.* (**12**), 6; 67 (Fig. 57)

The Plokiophilidæ are small anthocorid-like bugs and are divided into two subfamilies, namely, **Embiophilinæ** Carayon 1961 with one species *Embiophila myersi* China 1953 which was discovered in the webs of Embiidæ and **Plokiophilinæ** China 1953, with the species *Plokiophila cubana* (China and Myers) 1929, found in the web of a Theraposid spider on the underside of a log.

It is assumed that they are predaceous insects but hardly any reliable information is, at present, available.

It is of interest that while the ova of the **Embiophilinæ** do not develop before deposition, those of the **Plokiophilinæ,** already in the embryonic stage in the ovaries, always contain neanides almost completely formed at the moment of exit from the female.

The **Plokiophilinæ** were first considered to be a subfamily of the Microphysidæ but were later accorded family rank by Carayon in 1961. *Plokiophila* replaces *Arachnophila* China and Myers 1929, this name being preoccupied (Salvadori 1874; Aves).

The **Embiophilinæ** have the pronotum, scutellum and hemelytra dull, but the head, pronotal calli and anterior pronotal collar are shining. Very distinctly pilose, the hairs long, semi-erect, curved and directed posteriorly. Head moderately exserted; eyes relatively small, extending well on to the ventral side; ocelli large; tylus much shorter and broader than in *Plokiophila*; rostrum with third segment much shorter than fourth; metathoracic wing well-developed; venation as in *Plokiophila*; hamus completely absent; legs relatively short; femora incrassate, the anterior and median femora with short teeth; posterior femora with a row of bristles; no arolia or pseudoarolia distinguishable.

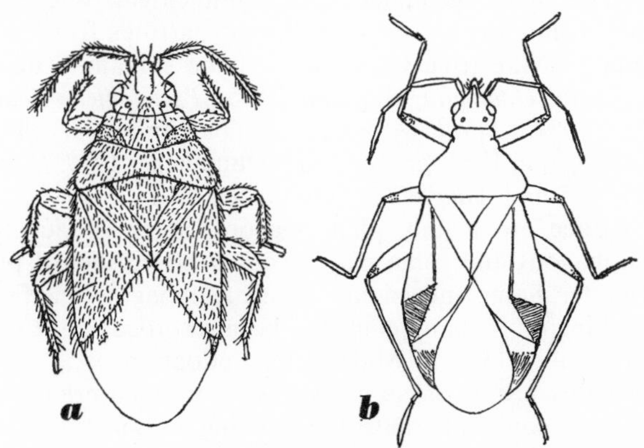

Fig. 57
A. *Embiophila myersi* China 1953; B. *Plokiophila cubana* (China and Myers) 1929 (after China) (Plokiophilidæ).

The principal characters of the **Plokiophilinæ** are: pronotum and scutellum dull, not shining, distinctly pilose; head more or less shining, strongly exserted, narrow, elongate. Rostrum very long, extending to posterior coxæ, apparently with three segments (the first segment concealed). Pronotum pubescent, strongly deflexed. Hemelytra well-developed, dull, covered with semi-erect hairs; cuneus distinct; membrane without veins, except for a short sclerotic thickening at base; metathoracic wings well-developed, narrow without hamus. Legs long, slender; tarsi very long with two segments. Abdomen laterally compressed; ovipositor absent; female genital orifice transverse.

References

Carayon 1961; China and Myers 1929; China 1953a.

MIRIDÆ (Hahn) 1831, *Wanz. Ins.* **1,** 234 (Figs. 46, 58)

This is a large family of mostly small insects with a delicate integument, sometimes glabrous or covered with fine and easily removed pubescence. They are variously coloured, but mostly pale. After the final ecdysis when the insect becomes adult the colour develops gradually.

The Miridæ have large eyes but no ocelli, antennæ composed of four segments and hemelytra with a distinct clavus, corium and usually a large cuneus; the membrane usually has two cells. Alary polymorphism often occurs. The legs are slender, as a rule, but in some genera, the femora are somewhat incrassate; the tarsi have three segments, the apical one bearing claws and arolia in most genera. The male genitalia are asymmetrical. Generally the scutellum is smooth, the most striking departures from this shape being found in the tribe Odoniellini of the subfamily **Bryocorinæ,** in the genera *Odoniella* Haglund 1895, *Pseudoniella* China and Carvalho 1951, *Parabryocoropsis* China and Carvalho 1951, *Distantiella* China 1944, *Bryocoropsis* Schumacher 1919, *Sahlbergella* Haglund 1895 and *Yangambia* Schouteden 1942.

The Miridæ are mainly phytophagous, some species, however, are oligophagous and some may favour a mixed diet of plant and animal matter. Some, indeed, are mainly zoophagous and facultative blood-sucking on human beings has been recorded, e.g. *Creontiades pallidifer* Walker 1873 in Malaya. The adoption of a carnivorous habit probably has its origin in the ease in which a primitive phytophagous bug could alternate the sucking of plant sap with the piercing of small insects associated with the same host-plant.

The Miridæ are diurnal but are occasionally attracted to light. Cannibalism has been observed in *Myrmecoris gracilis* Sahlberg 1920, when a male in attempting to copulate, inserted its stylets between the pro- and mesothorax of a female which was immediately affected by the saliva injected and soon died. The male then proceeded to suck the body fluids of the female.

The raptorial type of anterior leg is not represented in zoophagous Miridæ, but sometimes the anterior tibiæ are relatively more robust and somewhat incrassate apically, e.g. *Myrmecoris* Gorski 1852, *Systellonotus* Fieber 1858, *Cremnocephalus* Fieber 1860 and *Pilophorus* Westwood 1839. In some other genera no modification is to be seen. The median tibiæ may be modified in a similar manner to that of the anterior tibiæ.

Several of the phytophagous species are of economic importance, e.g. *Sahlbergella singularis* Haglund 1895 which attacks cacao in West Africa, *Kiambura coffeæ* China 1936, a pest of the coffee plant in East Africa and *Calocoris fulvomaculatus* de Geer 1773 which damages fruit trees in Europe. *Hyalopeplus similis* Poppius 1912 oviposits on tea shoots and also feeds on the flowers of beans used as a cover crop in Nyasaland, *Pleurochilophorus* sp. Reuter 1905 feeds on pigeon pea (*Cajanus indicus* Spreng.) and is also found on cotton (*Gossypium hirsutum* Linnæus), maize (*Zea mays* Linnæus), Sesamum (*Sesamum indicum* Linnæus), wild *Crotalaria* and *Gynandropsis* in Uganda. *Deraeocoris limbatus* Miller 1956 preys on Psyllids and *Campyloneura agalegæ* Miller 1956 is a predator of *Tetranychus* in the Agalega Islands.

Another economically important genus is *Helopeltis* Signoret 1858 (**Bryocorinæ**) species of which cause extensive damage to tea and cinchona in India, Ceylon, Malaysia and Africa. They also feed on many other plants of various orders, an important fact to take note of when control measures are under consideration. Some species are more or less restricted to their particular biotope, but some, notably *Helopeltis*, periodically migrate from the cultivated crop on which they have been feeding, to neighbouring wooded and non-cultivated areas.

Both neanides and adults of *Helopeltis* spp. attack the young leaves and shoots of the host-plant, the result being that the area around the site of the puncture from their stylets turns brown and finally black. In the event of severe attack the entire leaf shrivels, and in the case of the tea plant, is useless for the manufacture of tea.

Among zoophagous species may be mentioned *Deræocoris ruber* (Linnæus) 1758 (**Phylinæ**), predaceous on aphids, *Cyrtorhinus mundulus* (Breddin) 1896 (**Orthotylinæ**), an enemy of the sugar-cane pest *Perkinsiella saccharicida* Kirkaldy (Homoptera), *Stethoconus cyrtopeltis* Flor 1860 (**Phylinæ**) predaceous on *Stephanitis pyri* (Fabricius) 1803 (Tingidæ) and *Campyloneura virgula* (Herrich-Schaeffer) 1835 (**Phylinæ**) which feeds on Psocidæ.

Myrmecophilous Miridæ have been recorded, for example, *Lissocapsus wasmanni* Bergroth 1903 (**Phylinæ**) which lives in the nest of *Cremastogaster ranavalonis* Forel in Madagascar, *Systellonotus* *triguttatus* Linnæus 1758 has been found in company with the ants *Formica fusca* and *Lasius niger* and *Myrmecoris gracilis* Sahlberg 1920 with *Formica rufa*.

Recently an interesting case of the association of a carnivorous Mirid *Cyrtopeltis droseræ* China 1953 (**Phylinæ**) has been recorded. This Mirid has been found living on various species of Sundew (*Drosera*) in Western Australia. The extraordinary fact in this connection is the ability of this species to move freely over the sticky glandular hairs of the leaves without being entangled. Their food consists of freshly captured flies.

Another Mirid, *Setocoris bybliphilus* China and Carvalho 1951 (**Phylinæ**) has been found on the insectivorous plant *Byblis gigantea* also in Western Australia. When these Mirids walk on the leaves they rarely place more than two legs at a time on the sticky parts. In this way they are able to extricate themselves by means of the remaining legs, which are placed on the parts from which the sticky substance is absent.

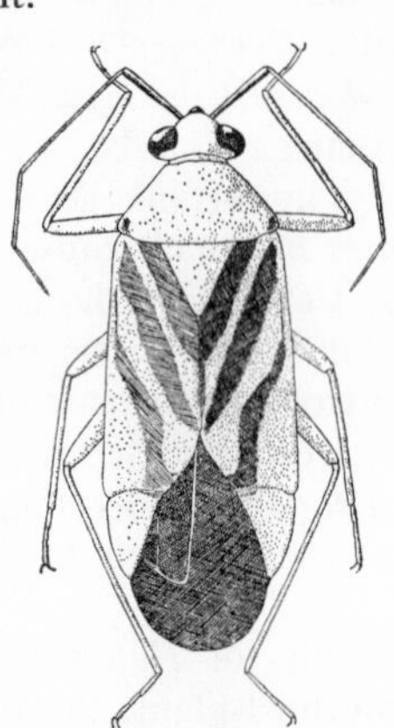

Fig. 58

Stenotus elegans Poppius 1912
(Miridæ-Mirinæ).

The ova of Miridæ are usually cylindrical and somewhat curved with a narrow differentiated portion of the chorion. Some ova have one or more slender processes through which air is able to enter. Examples of such ova are those of *Cyllocoris* Hahn 1834 (**Orthotylinæ**), *Helopeltis* and *Megacoelum* Fieber 1958 (**Mirinæ**). These ova are inserted into the tissue of the host-plant by the females.

Oviposition may be on the plant (the usual method) or inserted into the softer parts of the plant such as the young shoots. The fact that ova are inserted by the female into the young shoots is a considerable factor in control, especially in the case of *Helopeltis*, since those parts of the tea bush in which this pest oviposits, are systematically plucked for the preparation of the dried leaf.

Regarding eclosion, ecdysis and other incidents in the course of development of Miridæ, much information has been published on the European species, particularly by Küllenberg. This author also gives details of the various odours of the secretions, some of which, according to him, are by no means unpleasant to the human sense of smell.

There is not much information concerning the natural enemies of the Miridæ. They appear to be birds, among which the chaffinch, flycatchers and tits, spiders and predaceous Hemiptera.

The Miridæ are divided into the following subfamilies: **Mirinæ** (Amyot and Serville) 1843, elongate insects with the pronotal collar always present and well separated from the anterior pronotal lobe by a sulcus; arolia distinctly divergent towards their apices, usually dilated. Distribution, world-wide. **Orthotylinæ** Van Duzee 1916, with the arolia parallel or convergent towards apices, usually slender; pronotal collar, if present, of the depressed type, not separated from the anteriorlobe of the pronotum by a sulcus. Distribution, world-wide. **Bryocorinæ** (Douglas and Scott) 1865, with the pseudarolia arising from the ventral surface of the claw; hemelytral membrane with one cell; tarsi thickened towards apices. Distribution, holotropical. **Deræocorinæ** (Douglas and Scott) 1865, with the pseudarolia absent; pronotal collar present or if not, claws very long, smooth and slender; hemelytra opaque with two cells in the membrane; claws toothed or thickened basally. Distribution, world-wide. **Cylapinæ** Kirkaldy 1903, with the claws smooth basally, long and slender; pseudarolia absent. Distribution, tropical and subtropical regions.

References

Bergroth 1903; Butler 1923; Carayon 1960; Carvalho 1952; China 1951; China and Carvalho 1951; Douglas 1865; Douglas and Scott 1865; Fox-Wilson 1925; Gerin 1954; Hinton 1962; Johnson 1934; Kullenberg 1942, 1943, 1944, 1947; Leston 1952; Miller 1937, 1956; Mjôberg 1906; Odihambo 1958; Saunders 1909; Usinger 1939, 1946b.

ISOMETOPIDÆ Fieber 1860, *Wien ent, Monatschr.* **4,** 259 (Fig. 59)

Small, flattened insects very close to the Miridæ but nearly always with ocelli; tarsi with two segments; head short and strongly deflexed almost perpendicularly to body; second antennal segment always strongly modified, either thickened or dilated or both; posterior femora incrassate. They have the same genital structures and the behaviour of the spermatozoa after copulation is the same as in the Miridæ. They are able to jump.

Antennal structure in *Alcecoris globosus* Carvalho 1951 is remarkable in that the basal segment is thick and bears a moderately long spine and the second segment is elliptical and strongly globose.

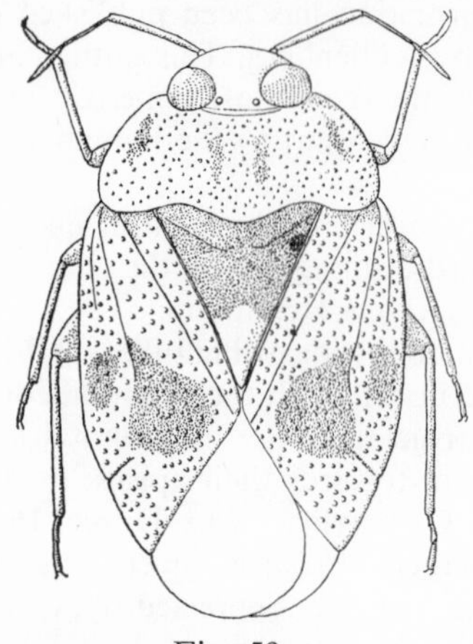

Fig. 59
Isometopus mirificus Mulsant and Rey 1897 (Isometopidæ).

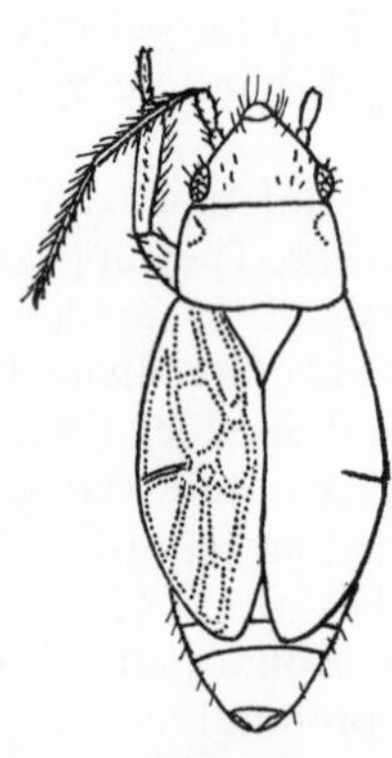

Fig. 60
Cryptostemma sordida China 1946 (after China) (Dipsocoridæ).

The very few known genera are distributed in the Palæarctic and Oriental Regions. Very little is known about their habits. *Letaba bedfordi* Hesse 1942, found in South Africa, is predaceous on the red scale (Coccidæ) of citrus.

References

Hesse 1947; Carvalho 1951; Carayon 1954c.

DIPSOCORIDÆ Dohrn 1859, *Cat. Hemipt.* 56 (Fig. 60)

This family contains very small dark-coloured insects which live in the soil among decaying leaves, under stones, in fact, any secluded spot that is not too dry. They are mostly nocturnal but sometimes appear in the daytime when they may be found on low vegetation and shrubs.

The site of oviposition is unknown but, judging by the habits of the adults, the ova are placed in the soil and among debris forming the usual habitat of the adult.

The only ova described so far are those extracted from dead females. They are relatively large, their length corresponding to that of the abdomen and their width indicates that not more than two ova can mature at the same time. The ornamentation of the ovum is very varied and complex.

The Dipsocoridæ are widely distributed.

References China 1946; Jordan 1940.

SCHIZOPTERIDÆ (Reuter) 1891, *Acta Soc. Sci. Fenn.* **19,** 3 (Figs. 46, 61)

The Schizopteridæ formerly placed in the Dipsocoridæ as a subfamily, are minute insects characterized by the very variable structure of the hemelytra, the strongly transverse head, which is deflexed and pressed between the anterior acetabula, the short rostrum, the antennæ inserted below the eyes and by having the pronotum transversely arcuately sulcate anteriorly. Brachypterous forms occur. Representatives live mainly in mosses and vegetable debris and are widely distributed.

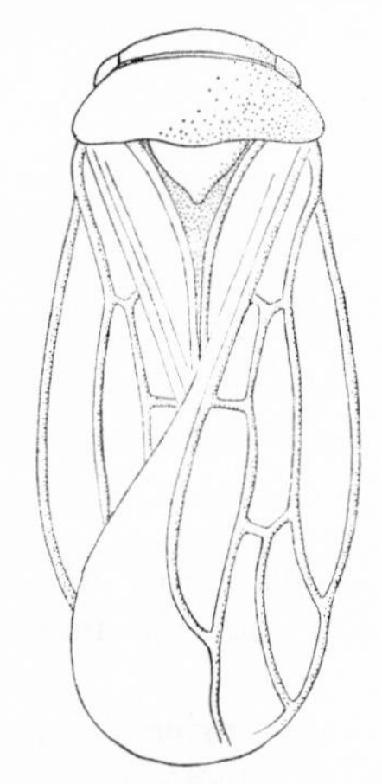

Fig. 61
Vilhennanus angolensis Wygodzinsky 1950 (Schizopteridæ) (after Wygodzinsky).

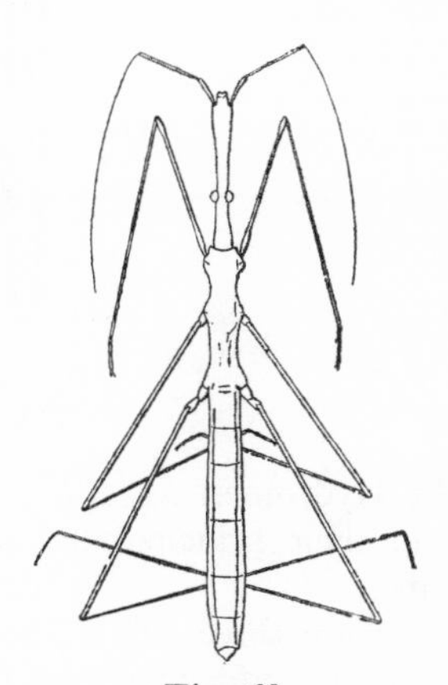

Fig. 62
Hydrometra stagnorum Linnæus 1758 (Hydrometridæ-Hydrometrinæ).

The ovum of *Vilhennanus angolensis* Wygodzinsky 1950 is shown in Fig. 46, 15.

References Wygodzinsky 1947a, 1950.

HYDROMETRIDÆ (Billberg) 1820, *Enum. Ins. Mus. Billb.* 67 (Figs 46, 62)

Long and very slender aquatic Heteropters with long, slender legs and antennæ. The head is elongate and much longer than the thorax. Ocelli are absent. The tarsi are composed of three segments and are provided with claws.

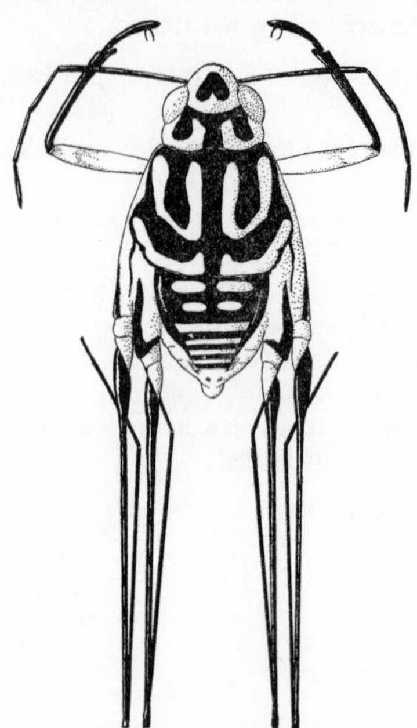

Fig. 63
Metrocoris ståli Dohrn 1860
(Gerridæ-Veliinæ)

The Hydrometridæ appear to subsist on dead prey which they find in their principal habitats, calm water and the margins of streams.

There are three subfamilies, namely, **Hydrometrinæ** Esaki 1927, in which the antennæ have four segments, the metasternum lacks an *omphalium*, and the claws are inserted apically on the tarsi. Distribution, world-wide; **Limnobatodinæ** Esaki 1927; in this the antennæ have five segments, the head and pronotum short spines,

---

Plate V (*facing*)
Belostomatidæ, Naucoridæ, Gerridæ, Nepidæ

1. *Lethocerus niloticum* (Stål) 1854. Belostomatidæ.
2. *Abedus ovatus* Stål 1862. Belostomatidæ.
3. *Cheirochela feana* Montandon 1897. Naucoridæ.
4. *Limnometra femorutu* Mayr 1865. Gerridæ.
5. *Cylindrostethus productus* Spinola 1840. Gerridæ.
6. *Laccotrephes fabricii* Stål 1863. Nepidæ.
7. *Ranatra elongata* (Fabricius) 1790. Nepidæ.

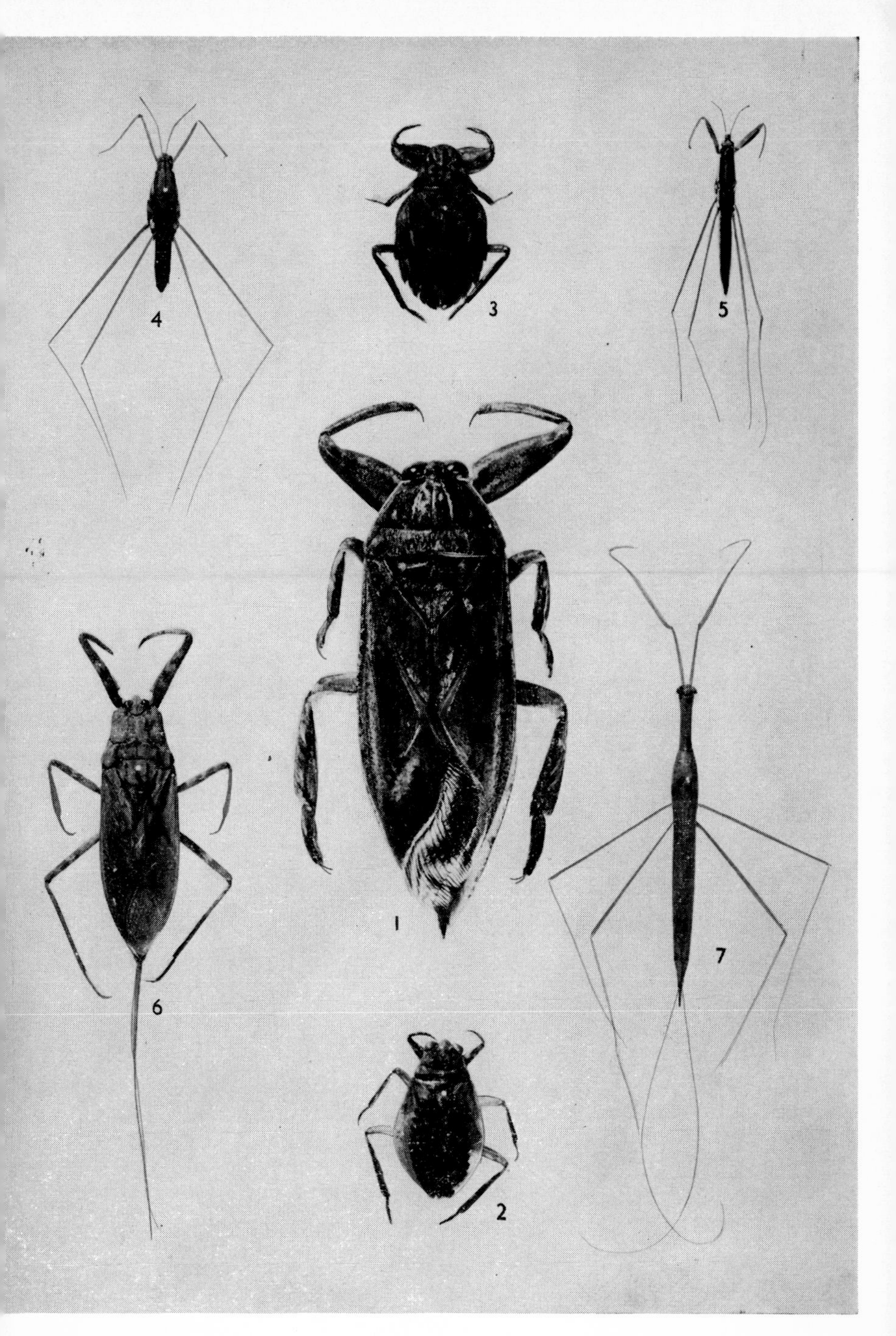

Plate V

the metasternum has a simple *omphalium* at the middle of the posterior margin, the head is sub-cylindrical behind the eyes and without ocellar spots, the posterior pair of trichobothrial setæ are inserted on small tubercles at extreme base of the head, the hemelytra have two closed cells and the small tarsal claws are inserted sub-apically dorsally. Distribution, Neotropical Region; **Hetero-cleptinæ** (=**Hydrobatodinæ** China and Usinger 1949), Villiers 1948 have antennæ with five segments, two ocellar spots behind the eyes, head and pronotum with short fine pilosity, posterior lobe of the head with posterior pair of trichobothrial setæ inserted on prominent rounded elevations, anterior lobe of head with two pairs of erect, stiff setæ, the metasternum with two separated *omphalia* on the posterior margin and the hemelytra with two closed cells. Distribution, Ethiopian Region.

Reference

Sprague 1956.

GERRIDÆ Leach 1815, *Brewster's Edinb. Encyc.* **9**, 123 (Plate V) (Figs. 46, 63)

Small or moderately large insects living on the surface of fresh, brackish or sea water. In moving over the surface of the water they use the median and posterior legs simultaneously and are able to swim against a strong current. The anterior legs, not of a predatory type, are used for seizing their food, which consists of living and dead organisms. Sexual dimorphism occurs. The adults have metathoracic glands.

The females oviposit on seaweed, floating feathers and even on the plumage of birds. Occasionally, oviposition occurs on an individual of the same species.

The species *Halobates sericeus* Eschscholtz 1822 which lives in the open ocean feeds on a small anemone and *H. hawaiiensis* Usinger 1938 feeds on any insect which may fall into the water and is strongly cannibalistic under laboratory conditions. *H. sericeus* is occasionally washed up on beaches during storms. *H. micans* Eschscholtz 1822 suffers a similar fate at times and is then preyed on by shore birds and crabs. Only one cosmopolitan species occurs in the Atlantic Ocean, while the Pacific Ocean abounds in species, with three or four species occurring in the open ocean, thousands of miles from land. The distribution of species appears to be determined mainly by water temperatures and ocean currents. The Philippine Islands and New Guinea apparently have the greatest number of species.

The genus *Asclepios* Distant 1915 is exclusively halophilous, and specimens, the bulk of which were *in copula* have been seen in enormous numbers on the surface of the sea along the beach.

Hermatobates haddonii Carpenter 1892, unlike other water-striders, inhabits coral reefs which are exposed for a short time only during ebb tide. At other times it submerges and hides probably in cracks in reefs or under dead shell or reef blocks, on the surface of which it can climb or walk rapidly. *Rheumatobates crinitus* Herring 1949 is the only known marine species of the genus. It has been found in mangrove swamps and in the open bay of Florida Keys.

The Gerridæ contains five subfamilies: **Gerrinæ** Bianchi 1896 with an elongate body, the abdomen elongate, more than twice as long as wide at base, the internal margin of the eyes concavely emarginate, the antennæ moderately long. Distribution, world-wide. **Halobatinæ** Bianchi 1896 small, stumpy insects with the internal margin of the eyes convex, the anterior legs thickened, the abdomen usually small and extending distinctly beyond the apices of the posterior coxæ, the segments not fused, the male genital segments backwardly porrect, not folded ventrally with the apex directed towards the head. Distribution, Atlantic, Indian and Pacific Oceans. **Ptilomerinæ** Esaki 1927, have a cylindrical body, prominent eyes with the internal margin distinctly emarginate, long, slender antennæ which are longer than half the length of the body, median and posterior legs with tibiæ and tarsi very long and slender, thread-like, the tarsi with one very short segment, and the median tibiæ of the male with a fringe of long, thin hairs. The **Ptilomerinæ** are medium to large water-skaters living on the surface of running water. Distribution, Indo-Australian Region. **Hermatobatinæ** Coutière and Martin 1901, minute marine insects with an ovate, stumpy body entirely covered with fine pubescence, with the abdomen hardly extending beyond the apices of the posterior coxæ, the segments fused and indicated only by the presence of spiracles at the sides, the meso- and metanotum fused, the male genital segments folded ventrally to lie with the apex directed towards the head and the legs short and relatively robust. Distribution, Australian Region. **Rhagadotarsinæ** Lundblad 1933, have a wide, short head produced in front of the eyes, very large eyes with the internal margin not concave, the apical segments cylindrical, elongate and long anterior femora. Distribution, Oriental and South Ethiopian Regions.

References

Buchanan White 1883; Esaki 1924, 1930; Herring 1949; Lundbeck 1914; Myers 1926; Parshley 1917; Tonapi 1959; Usinger 1938, 1951; Usinger and Herring 1956.

VELIIDÆ (Amyot and Serville), 1843, *Hist. Nat. Hémipt.*, **1**, 418 (Fig. 64)

Mostly small insects with the body widened at the level of the prothorax. There are often two forms, one macropterous and the other apterous. The legs are inserted more or less equidistantly (except in **Haloveliinæ**), the vertex usually has a distinct, percurrent, median, longitudinal suture or glabrous line, nearly obsolete, in which case the eyes are small and do not extend backwards on to the sides of the pronotum; the odoriferous glands usually have paired lateral channels terminating above the posterior acetabula in a tuft of hairs; inner margins of eyes straight; male harpagones large and distinct.

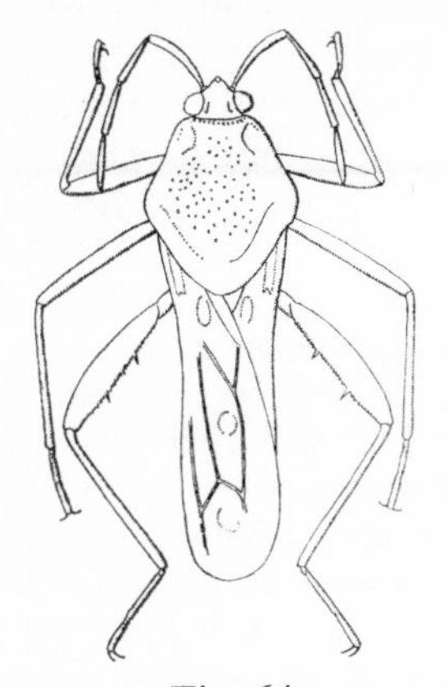

Fig. 64

Velia africana Tamanini 1946 (Veliidæ-Veliinæ).

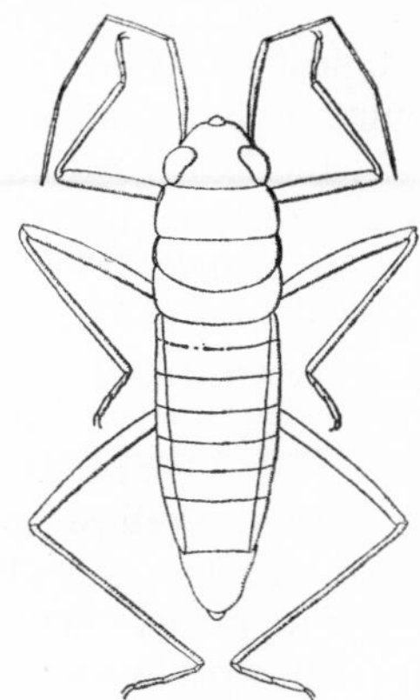

Fig. 65

Mesovelia furcata Mulsant and Rey 1852 (Mesoveliidæ).

The Veliidæ are gregarious and predaceous. In movement they use only their median legs. In *Rhagovelia* Mayr 1865 (**Rhagoveliinæ**) the apical segment of the tarsus of the median legs is furcate and has a tuft of feathery setæ which spread out on the water during movement. The tarsi of *Tetraripis* Lundblad 1936 (**Rhagoveliinæ**) are notched and have plumose setæ like the median tarsi.

When Veliidæ congregate, as they do in summer, they will be seen to disperse momentarily if disturbed, but to reunite soon afterwards. The reason for such congregations is probably mutual attraction.

A marine Veliid, *Halovelia mariannarum* Usinger 1946, in Samoa lives along the edges of a shallow lagoon and is closely associated with volcanic rocks which occur in the intertidal zone. It is active at mean tide, but hides in rocks at low and high tide. Its food is mainly midges and Collembola. The curious halophilous water-strider *Halovelia maritima* Bergroth 1924 is recorded as being abundant on the surface of sea water among rocks.

There are six subfamilies of Veliidæ; **Perittopinæ** China and Usinger 1949, with the basal segments of the tarsi of anterior and posterior legs very short and inconspicuous, of the median legs long, subequal to the apical segments; hemelytra divided into corium and membrane, the corium with two closed cells; membrane broad and long without veins; tarsal formula 2 : 3 : 3. Distribution, India and Malay Archipelago. **Rhagoveliinæ** China and Usinger 1949, have the hemelytra undivided into a distinct corium and membrane, the median tarsi deeply cleft with laminate claws and plumose hairs arising from the base of the cleft. Distribution, Holotropical and sub-tropical. **Hebroveliinæ** Lundblad 1939. In this subfamily, the tarsal claws are all terminal, the tarsal formula 1 : 2 : 2; the hemelytra have five closed cells, one of which along the costal margin. Distribution, Ethiopian Region; **Microveliinæ** China and Usinger 1949. In this subfamily the tarsal claws are preapical, the tarsal formula 1 : 2 : 2, and the hemelytra have four closed cells. Distribution, world-wide. **Haloveliinæ** Esaki 1930, are very small, sub-oval and usually apterous; macropterous forms are unknown. The tarsal formula is 2 : 2 : 2. Distribution, Indian and Pacific Oceans. **Veliinæ** China and Usinger 1949 are elongate, usually macropterous with the mesonotum largely concealed by the pronotum in macropterous forms. The tarsal formula is 3 : 3 : 3. Distribution, world-wide.

There is little information about the developmental stages of Veliidæ.

References Esaki 1924; Jordan 1932; Kellen 1959.

MESOVELIIDÆ Douglas and Scott 1867, *Ent. Mon. Mag.* **4,** 3 (Figs. 46, 65)

A small family containing small, slender macropterous or apterous insects of dull colouration. They have a wide head with moderately large eyes, ocelli, a long rostrum and long, slender antennæ. The tarsi have three segments, claws but no arolia.

As regards habitat, they prefer pieces of smooth water with an abundant vegetation on which they walk and feed on dead and feeble organisms. Some live in moss on bark of trees near water or on floating vegetation. A cavernicolous and halophilous species *Speovelia maritima* Esaki 1929, has been found living on the walls of a cave communicating with the sea in Japan, in which light is totally absent.

The females insert their ova into vegetable tissue.

The ovum of *Mesovelia mulsanti* Buchanan White 1897 is shown in Fig. 46, 28.

There are three subfamilies, namely **Mesoveloideinæ** Usinger (in China and Miller 1959), in which the head is short and strongly declivous, the antennæ are inserted close to the anterior margin of the eyes, the posterior margin of the eyes is contiguous with the anterior pronotal margin and the scutellum is simple, triangular, somewhat shorter than broad at base. Distribution, South America. **Macroveliinæ** McKinstry 1942, which have the pronotum extending posteriorly to cover the scutellum which is apparently absent, the hemelytra with the corium not differentiated from the membrane, the whole with six closed cells. Distribution, California, U.S.A. **Mesoveliinæ** Douglas and Scott 1867, a small subfamily in which

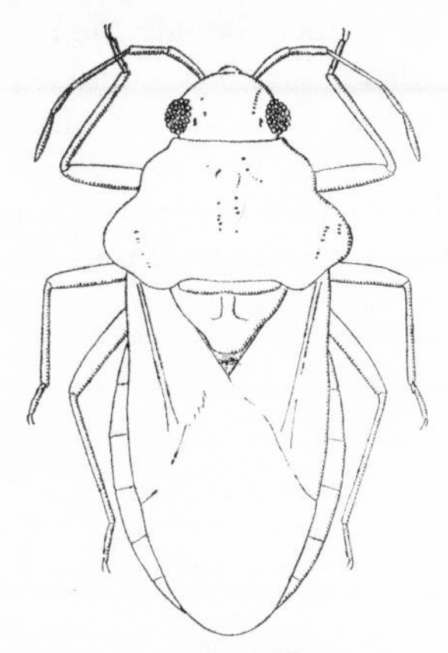

Fig. 66

Hebrus pusillus Fallèn 1807 (Hebridæ).

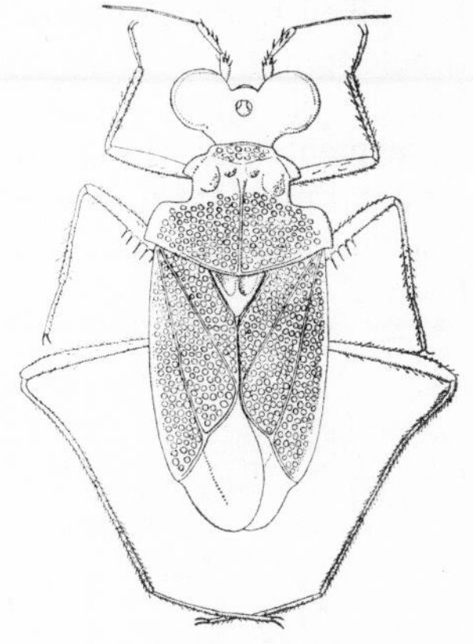

Fig. 67

Leotichius speluncarum China 1941 (Leotichiidæ).

the pronotum does not extend posteriorly, the scutellum is distinctly visible and consists of two lobes, the basal one convex, the apical one spatulate, the membrane of the hemelytron without cells, the ocelli present in macropterous form or absent in apterous form, which is common. Distribution, world-wide.

Reference Yuasa 1929.

HEBRIDÆ (Amyot and Serville) 1843, *Hist. Nat. Hémipt.* xl, 293 (Fig. 66)

Very small, aquatic insects with a short, thick body, large eyes, ocelli, a long rostrum, apparently with three segments which lies in a distinct longitudinal groove formed by the bucculæ when not in use, antennæ composed of four or five segments, a small scutellum, hemelytra with the corium with one cell and the clavus membranous, wide and without nervures. The legs are short and robust and similar to each other. Odoriferous glands are present. The Hebridæ live in swamps and on moss on banks of streams.

They are distributed in the Palærctic, Nearctic, Neotropical, Ethiopian and Oriental Regions.

LEOTICHIIDÆ China 1933, *Ann. Mag. Nat. Hist.* (10) **12,** 185 (Fig. 67)

This family contains a single genus *Leotichius* Distant 1904, found, so far as is known, only in the Indo-Malayan Region.

It is allied to the Leptopodidæ but it differs in having the rostrum without spines, short antennæ of which the basal and second segments are thick, a carinate pronotum, the anterior tarsi with one segment and the remaining tarsi with two segments. The compound eyes are on the ventral surface of the head and are directed downwards and the dorsal surface of the head is densely covered with short setæ.

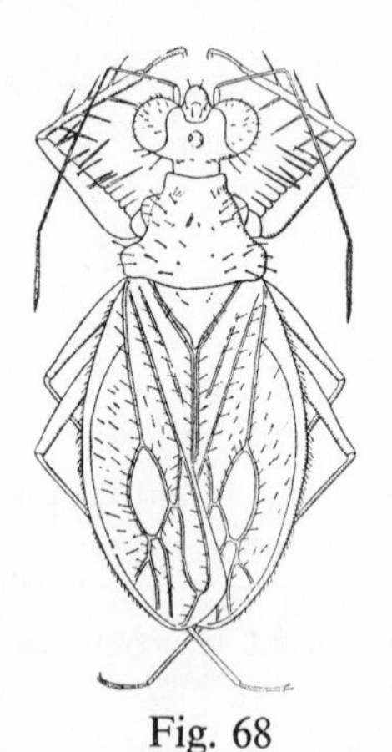

Fig. 68

Leptopus marmoratus Horvath 1897 (Leptopodidæ).

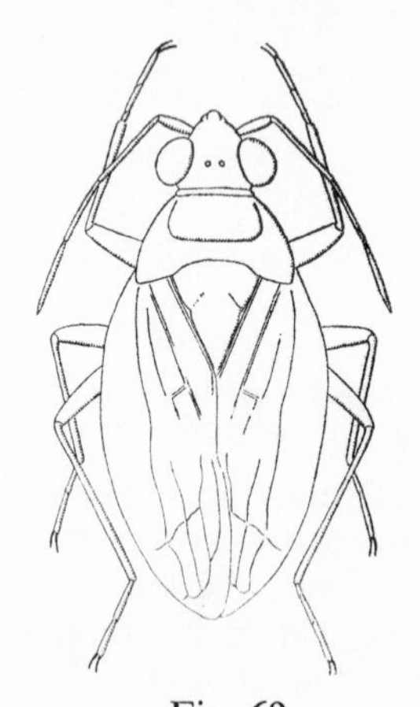

Fig. 69

Salda littoralis Linnæus 1758 (Saldidæ-Saldinæ).

One species, *Leotichius speluncarum* China 1941 has been found on bat guano in limestone caves in Malaya and *L. glaucopis* Distant 1904 in caves in Burma.

LEPTOPODIDÆ Costa 1838, *Cimic. Regni Neap. Cent.* **1,** 151 (Fig. 68)

Small, dull-coloured insects allied to the Saldidæ. They are characterized by the short rostrum, the basal segment of which (or sometimes segments one and two) are spinose, by prominent ocelli, subpedunculate eyes, incrassate anterior femora and tibiæ armed with long, slender spines.

So far as is known, Leptopodidæ are confined almost entirely to the tropical and subtropical regions of the eastern hemisphere, but do not appear to be abundant, possibly because they are overlooked on account of their small size. *Valleriola wilsonæ* Drake 1956 has been recorded from Australia for the first time. Little is known about their ecology.

Neanides have only one abdominal scent gland, the ostiole of which is located between segments three and four.

References

Drake 1956; Drake and Hoberlandt 1950.

SALDIDÆ (Amyot and Serville), 1843, *Hist. Nat. Hémipt.* xlix (Figs. 46, 69)

Generally dark-coloured, soft-bodied insects found in proximity to fresh water streams. They probably feed on insect remains and dipterous larvæ, but this is uncertain. Some species are halophilous. They are able to run and fly off when disturbed.

Saldidæ are mostly oval shaped in outline with a short, wide head and prominent eyes. The antennæ are long, also the rostrum, which does not lie on the ventral surface of the head when not in use. The eyes are large with the inner margins excavate; ocelli are present and often contiguous; the hemelytral membrane has four, rarely five parallel-sided closed cells; the mandibular plates are prominently convex, transverse and shining; the scutellum is large and triangular, usually longer than wide. Apterous forms are unknown. Neanides have one abdominal scent gland, the ostiole of which is situated between segments three and four.

The Saldidæ have been considered to be the most primitive of living Heteroptera in respect of their short gular area, the ventrally directed mouth, their feeding habits, habitat and methods of oviposition. This opinion, however, has not been generally accepted.

Aepophilus bonnairei Signoret 1879 (**Aepophilinæ**) (Fig. 70) is a small ovate, delicate insect which lives on the shore at the tidal zone and conceals itself under deeply embedded stones at high tide. When the tide is low it moves about on stranded seaweed and is often in association with Coleoptera. The neanides are gregarious.

They have only one abdominal odoriferous gland which is situated between segments three and four. *A. bonnairei* is to a large extent submarine, the adults and neanides apparently preying on small Crustacea.

Lampracantha crassicornis (Uhler) 1877 (Saldinæ) has been found on muddy ground in company with *Ochterus americanus* Uhler 1877, the majority of specimens on matted marsh grass and sedge growing in water of varying depths.

Halosalda lateralis Fallèn 1807 (Saldinæ), occurs in salt marshes just above the high tide mark and *Saldula pallipes* (Fabricius) 1794 (Saldinæ) extends far into the intertidal zone. At high tide both adults and neanides (especially the latter) of *S. pallipes* have been observed to remain at the lower levels and to be submerged by the rising tide, instead of retreating before it.

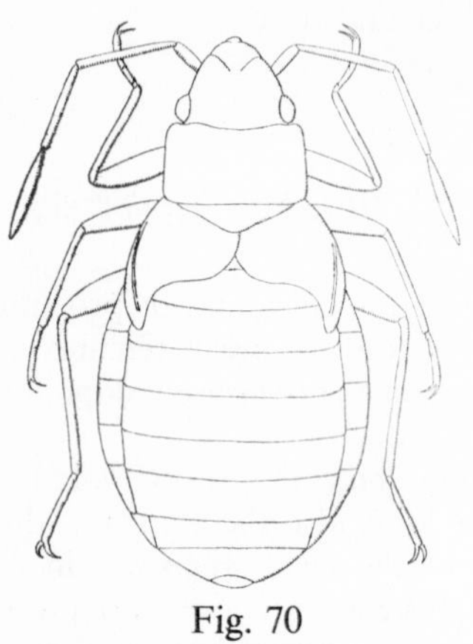

Fig. 70
Aepophilus bonnairei Signoret 1897 (Saldidæ-Aepophilinæ).

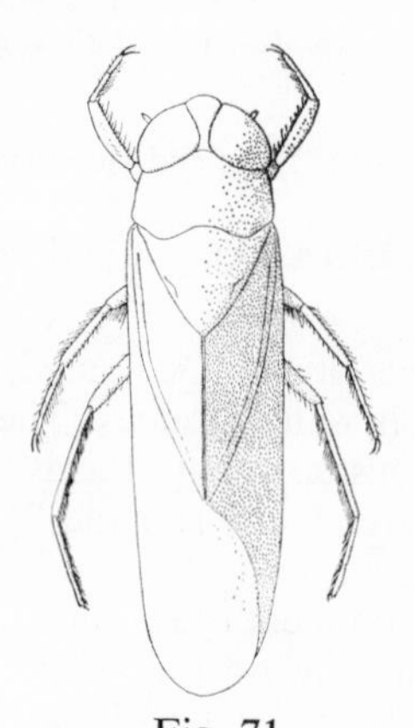

Fig. 71
Anisops sardea Herrich-Schaeffer 1849 (Notonectidæ).

The Saldidæ contains three subfamilies: **Aepophilinæ** (Puton) 1879, in which the scutellum is transverse, much wider than long, narrowly arcuate, the eyes small with the inner margins entire, ocelli absent, the pronotum quadrate and the hemelytra always abbreviated with the apical margins concave and the outer angle produced. Distribution, South-West Europe. **Saldoidinæ** Reuter 1912, in which the scutellum is large, triangular, at least as long as wide, the eyes large, usually with the inner margin sinuate, ocelli are present and often placed close together, the hemelytra are never abbreviated with the apical margin concave and the pronotum is more or less trapezoidal with two erect tubercles or spinous processes. The inner margins of the eyes are parallel, not emarginate. Distribution,

Florida, U.S.A., Formosa and the Philippine Islands. **Saldinæ** Van Duzee 1917, has similar characters to those of the previously mentioned subfamily, but the pronotum lacks tubercles or spinous processes, the head is wide and the inner margins of the eyes are usually marginate. Distribution, world-wide.

References

Baudoin 1939; Bergroth 1899; Brindley 1934; Brown 1948; China 1927; Cobben 1957; Ekblom 1926; Jordan and Wendt 1938; Kellen 1960; Slater 1955; Spooner 1938; Wiley 1922.

NOTONECTIDÆ Leach 1815, *Brewster's Edinb. Encyc*, **9,** 124 (Figs. 46, 71)

The Notonectidæ are elongate, wedge-shaped, back-swimming insects with air-bubble respiration. The abdomen lacks a midventral carina, the antennæ usually have four segments, the posterior legs are long, oar-like, formed for swimming, the eyes large, ocelli absent, hemelytra without venation, scutellum triangular, large. Odoriferous glands are present.

Larger species of Notonectidæ attack the fry of fish, young batrachians, while smaller species and neanides attack small crustacea. The males are able to stridulate.

The female, when ovipositing, either inserts the ova into vegetable tissue or affixes them to a plant-stem or other object (Fig. 71, 30).

There are two subfamilies: **Notonectinæ** Leach 1815, in which the hemelytral commissure lacks a pit (sensory organ), at the anterior end, the median femora have a sub-apical spur on the posterior margin, the antennæ have four well-defined segments and the male genitalia are symmetrical. Distribution, world-wide. **Anisopinæ** Hutchinson 1929, in which a pit is present, the median femora lack a sub-apical spur and the antennæ have three well-defined segments. Distribution, tropics and sub-tropics.

References Hale 1924; Larsen 1930; Walton 1936.

PLEIDÆ (Fieber) 1851, *Genera Hydroc.* 27 (Fig. 72)

Small insects which are remarkable in having the head and thorax partially fused. The form is oval and strongly convex, the abdomen has a distinct mid-ventral carina, the antennæ have three segments, the legs are formed for locomotion and the rostrum has four segments.

The habitat of Pleidæ is principally the still waters of ponds and lakes where they live among aquatic vegetation. They swim on the back and propel themselves by the posterior legs, the tibiæ and tarsi of which have two rows of setæ. They move from place to place often by clinging to floating vegetation. The females insert their ova into plants (Fig. 46, 26).

The Pleidæ are to be found in the Palæarctic, Nearctic, Oriental, Australian and Neotropical Regions.

Reference Wefelscheid 1912.

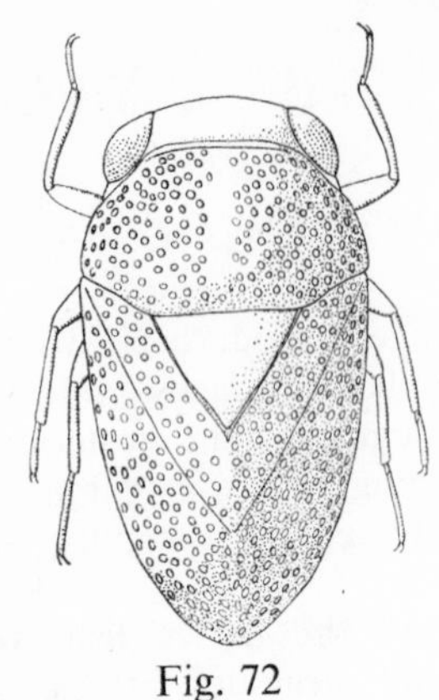

Fig. 72

Plea pullula Stål 1855 (Pleidæ).

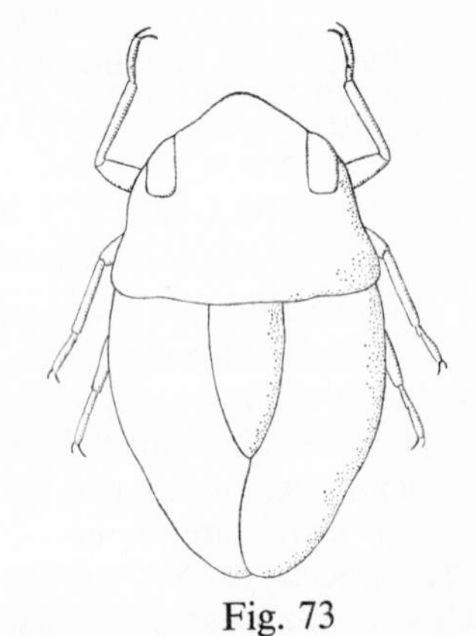

Fig. 73

Idiocoris lithophilus Esaki and China 1927 (Helotrephidæ-Neotrephinæ).

HELOTREPHIDÆ Esaki and China 1927,*Trans. R. ent. Soc. Lond.* 280 (Fig. 73)

The Helotrephidæ are allied to the Pleidæ which they resemble in having the head and pronotum fused. They have antennæ with one or two segments, a long scutellum and the male genital segments twisted to the left. The tarsi have spiniform arolia and a tubular membranous empodium. Very little is known about their ecology.

There are three subfamilies: **Neotrephinæ** China 1940 with the scutellum much longer than wide, the dorsal surface of the head separated from the ventral surface by a distinct, fine carina. The female has a sub-ovipositor. Habitat, pot-holes in mountain streams. Distribution, South America. **Idiocorinæ** Esaki and China 1927, with a flattened body, no suture between head and pronotum, the antennæ with one plate-like segment and the female without an ovipositor. Distribution, Tanganyika Territory in the lake of that name. **Helotrephinæ** Esaki and China 1927, with a strongly convex body, a suture between head and pronotum, the antennæ with two segments and the female without an ovipositor. Distribution, still water and lakes in the Oriental and Ethiopian Regions.

Reference Miyamoto, 1953.

CORIXIDÆ Leach 1815, *Brewster's Edinb. Encyc.* **9,** 124 (Fig. 74)

Moderately large or small aquatic insects usually dull in colour with the pronotum and hemelytra sometimes with a linear or vermiculate pattern. The very mobile and wide head has large eyes and a very short, broad and apparently unsegmented rostrum. The hemelytra and wings are well-developed. Scent glands are present in the neanides on the third, fourth and fifth segments of the abdomen. In the adult the glands are found on the metathorax with the ostioles at the sides of the median coxæ.

The anterior legs are short with the anterior tarsi modified into spatulate palæ margined with robust setæ (Fig. 4, 8). The median legs are long and slender and the posterior legs have robust femora, short tibiæ and flattened tarsi with marginal setæ. Each leg is adapted to a particular function; the anterior legs for collecting food, the median legs for supporting the animal when stationary and the posterior legs for swimming. Corixidæ swim with the ventral surface below. They are not confined to fresh water but also occur in brackish and salt water. *Trichocorixa wallengreni* (Stål) 1859, *reticulata* (Guérin) 1857 and *verticalis* Fieber 1851 occur in brackish water. It has been suggested that since *wallengreni* can support salinities considerably higher than those of the sea, its presence in California, Hawaii and China might conceivably be explained on the basis of travel by the North Equatorial current.

In order to breathe, the Corixid applies the side of the thorax to the surface of the water and by repeatedly bending its head drives air into the cavities situated between the head and prosternum and between the pro- and mesosternum. Finally, the air thus entrapped reaches the abdominal spiracles and the trachaeal system.

They are largely phytophagous, a fact which refutes the common assumption that all aquatic bugs, including those belonging to this family, are predaceous. Some species prefer waters in which there is a flourishing vegetation, while others prefer areas in which the vegetation is scanty. Breeding takes place sometimes in permanent water, sometimes in temporary pools.

The food of Corixidæ, according to Hungerford, is derived from the organic ooze on the bottom of a pond or lake. In gathering it they ingest other matter such as unicellular algæ, filaments of *Oscillatoria*, *Zygnema* and *Spirogyra* from which they suck the chlorophyll. In their food habits they have an advantage over all other families of strictly predaceous aquatic Heteroptera on account of the continued abundance of the food supply. While many species use the modified anterior tarsi or palæ for gathering their food, there are indeed Corixidæ in which these appendages are not modified, which suggests that they live by predation.

Several cases of predatory behaviour of Corixidæ have been observed both in the laboratory and in the field. The species concerned were, *Corixa* (*Corixa*) *punctata* (Illiger) 1807, *C.* (*Hesperocorixa*) *linnæi* (Fieber) 1848 and *Sigara* (*Subsigara*) *falleni* (Fieber) 1848 and occasionally unidentified adults and neanides. The prey comprised larvæ of Tendipedidæ, Culicidæ, Ephemeroptera and Tubificidæ. Neanides of Corixidæ have been seen feeding in the field on *Cladocera*.

Copulation takes place in the water and the ova are usually deposited on some submerged part of an aquatic plant, but a departure from this mode of oviposition is exhibited by the American species, *Rhamphocorixa balanodis* Abbott 1912 (**Corixinæ**) an inhabitant of muddy ponds. This species has been recorded as ovipositing on the carapace of a crayfish, *Cambarus immunis* Hagen, but whether this is the sole site chosen by the females on every occasion seems open to doubt. So far as is known, most Corixid ova are more or less oval with the apex sometimes sub-acutely conical, with or without a short pedical. The structure of Corixid ova has been described by Poisson. At the time of eclosion the embryo ruptures the apex of the chorion (Fig. 46, 18–21).

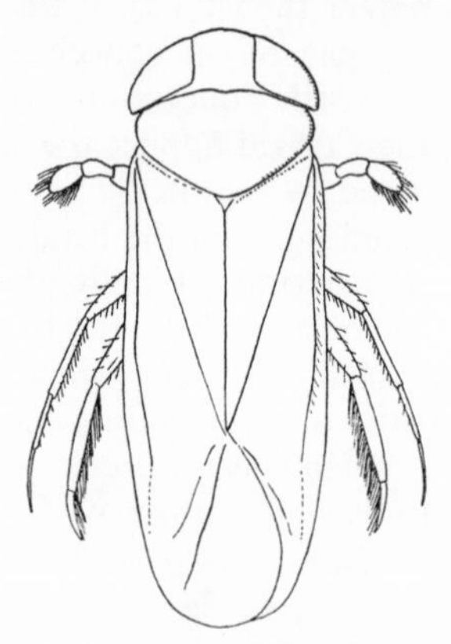

Fig. 74
Sigara pectoralis Fieber 1851 (Corixidæ).

The genus Corixa Geoffroy 1762 possesses stridulatory organs. The Corixidæ generally have well-developed wings, but in some genera, namely *Palmacorixa* Abbott 1913, *Krizousacorixa* Hungerford 1930 (**Corixinæ**) and *Cymatia* Flor 1860 (**Cymatiinæ**) the metathoracic wings are often reduced and non-functional. Two abundant species in Mexico, *Corixa mercenaria* Say 1931 and *Krizousacorixa femorata* (Hungerford) 1948 (**Corixinæ**) are collected and exported for food for song-birds, poultry and fish. The ova are collected by natives who make a kind of bread from them.

The Corixidæ, which are distributed in all zoogeographical regions, are divided into six subfamilies: **Diaprepocorinæ** Lundblad 1928 with the scutellum exposed and covered by the pronotum only at the anterior margin, the antennæ composed of four segments, ocelli present and the anterior claws large. The presence of ocelli indicates that they live in temporary pools of water; **Micronectinæ** Jaczewski 1924, in which the ocelli are absent, the antennæ have three segments and the anterior claws in the female spine-like, in the male flattened and capable of being folded back into an excavation on the outside of the tarsus; **Stenocorixinæ** Hungerford 1948, in which the hemelytra have no embolar groove, and the scutellum is concealed by the pronotum, its apex being rarely visible; **Cymatiinæ** Walton 1940, in which the hemelytra have an embolar groove, a nodal suture towards the apex of the embolium is absent and the rostrum lacks transverse sulcations; **Heterocorixinæ** Hungerford 1948, in which the rostrum has transverse sulcations, the scutellum is concealed by the pronotum, a nodal furrow is present, vein M usually curving upwards to fuse with Cu at, or just before the origin of the nodal furrow, the infra-ocular portion of the genæ very broad, the lower margin of the eye concave and vein M of the hemelytra is indistinct, parallel and very near to Cu; **Corixinæ** Douglas and Scott 1865, in which the infra-ocular portion of the genæ usually is not broad, but if so, the hypo-ocular suture, if present, arises about midway along the ventral margin of the eye, ocelli lacking and the scutellum is concealed by the pronotum.

References

Abbott 1912; Banks 1938; Davis 1964, 1965; Fernando and Leong 1963; Frost and Macan 1948; Griffith 1945; Hale 1924; Hungerford 1923, 1948; Hutchinson 1931; Jaczewski 1961; Jordan 1937; Kirkaldy 1901; Mitis 1935; Walton 1962; Westwood 1871.

NEPIDÆ (Plate V) (Latreille) 1802, *Hist. Nat. Crust. Ins.* **3**, 252

Moderately large or large insects, elongate or flattened and with raptorial legs. They have a small head, more or less horizontal, with prominent, globular eyes, a short rostrum composed of three segments, directed outwards and forwards. The abdomen has a pair of long, slender, posterior appendages forming a respiratory syphon. The posterior coxæ are short, free, rotatory. The tarsi are composed of one segment. Ocelli are absent. The metathoracic wings, though present, are reticulately veined, but not functional. Odoriferous glands appear to be absent from both neanides and adults. Hydrostatic organs are present on the abdomen which enable the possessors to remain within syphon reach of the surface film.

The Nepidæ contains two subfamilies: **Nepinæ** Douglas and Scott 1865, with an ovate, flattened body, a broad, trapezoidal pronotum, the anterior coxæ short, the anterior femora strongly incrassate and the head across the eyes distinctly narrower than the anterior margin of the pronotum; **Ranatrinæ** Douglas and Scott, 1865, elongate, cylindrical, the pronotum widened posteriorly, the head across the eyes distinctly wider than the anterior margin of the pronotum, the anterior coxæ long and slender and the anterior femora not strongly incrassate.

Although the distribution of the Nepidæ is wide, the majority of species is found in the Oriental, Ethiopian and Neotropical Regions.

Reference Davis 1961, 1964.

BELOSTOMATIDÆ (Leach) 1815, *Brewster's Edinb. Encyc.*, **9**, 23 (*Plate* 5)

This family contains large to fairly small aquatic and predaceous insects. They are somewhat flattened dorsally and convex ventrally. The abdomen is terminated by two retractile appendages formed from the eighth segment which, when joined, serve as a respiratory syphon. The anterior trochanters, femora, tibiæ and tarsi have very dense short setæ on the under surface (Fig. 4, 6). This setal clothing is also present on the remaining legs but is not equally dense. The median and posterior legs have a double fringe of fine, mostly long setæ. The rostrum is robust and the antennæ have four segments.

Some of the largest Heteroptera are to be found in this family, namely *Belostoma* Latreille 1807, *Lethocerus* Mayr 1852 and *Hydrocyrius* Spinola 1850. Although aquatic, representatives of the Belostomatidæ often fly from one pond or lake to another and during flight are frequently attracted to artificial light.

The Belostomatidæ are very voracious feeders and attack small fish, immature batrachians and molluscs. There is one recorded instance of a fairly large bird being attacked by *Hydrocyrius columbiæ* Spinola 1850, the bird being a Siberian ringed plover. It was found with the bug under its wing and was in a moribund condition. It was eventually killed for examination, which revealed that the bug had been feeding on its liver.

Females oviposit either on vegetation or on the backs of males. The species which have been recorded as ovipositing on the males belong to the genera *Abedus* Stål 1862, *Belostoma*, *Sphærodema* Laporte 1832, *Poissonia* Brown 1948 and *Hydrocyrius*, but whether this method of oviposition is invariable has yet to be confirmed;

also whether it is always the male and not another female which is the recipient of the ova (Fig. 46, 17).

Lethocerus indicum (Lepeletier and Serville) 1825 places its ova on vegetation in groups of several hundreds, sometimes. This species is eaten by the Laos people of Indo-China. The Belostomatidæ are widely distributed.

References

Chen and Young 1943; Game Dept. Uganda 1949; Lanck 1959; Severin and Severin 1910; Slater 1899.

NAUCORIDÆ Fallèn 1814, *Spec. Nov. Disp. Meth.* **3**, 15 (Plate V) (Fig. 75)

Moderately large aquatic insects of ovate form with a short rostrum and the hemelytral membrane lacking nervures. The anterior legs are raptorial. The outer margin of the eyes is continuous with the lateral margin of the dilated pronotum and front margin of the head. The apex of the abdomen lacks appendages. The median and posterior tibiæ and tarsi are cylindrical with rows of bristles and feeble setæ. Many species exhibit alary polymorphism. Respiration is of the plastron type.

They swim freely but often congregate among aquatic vegetation. The adults are able to stridulate and are also able to eject fluid from the rectum when disturbed. Their bite is painful.

Aphelochirus æstivalis (Fabricius) 1803 frequents running water in which stones and aquatic plants are found. It lives under them and to them the female attaches its ova.

The family is divided into eight subfamilies: **Naucorinæ** Stål 1876, with the anterior tarsi composed of one segment. The anterior margin of the pronotum is straight in the middle or feebly convex behind the interocular region. The internal margin of the eyes converges anteriorly. The gula is moderately long and carinate, the meso- and metasternum not, or feebly, carinate foveolate. **Limnocorinæ** Stål 1876, with the anterior tarsi having one segment. The pronotum and gula are similar to those of the **Naucorinæ,** but the internal margin of the eyes diverges anteriorly, the meso- and metasternum are strongly carinate with the apex deeply foveolate or broadly sulcate; **Laccocorinæ** Stål 1876 have the anterior tarsi composed of two segments with two claws which are often inconspicuous, the anterior margin of the head strongly turned downwards and backwards, the rostrum thus arising well behind the anterior margin of the head. The gular region is very short and non-carinate, the median and posterior femora each have two longitudinal rows of conspicuous bristles on the lower surface in addition to the usual two rows of short setæ along the posterior surface; **Cryphocricinæ**

Montandon 1897 with a flattened body and the rostrum articulated to the anterior margin of the head. Labrum well-developed. Head retracted into prothorax. Sternites of the abdomen not pubescent. Members of this subfamily are usually brachypterous. **Ambrysinæ** Usinger 1941. Anterior margin of the pronotum as in **Cryphocricinæ.** The body is ovate and the abdominal sternites densely pubescent. The rostrum is short and articulated to the anterior margin of the head. The carina on the gula and prosternum are of equal height. Always macropterous. **Cheirochelinæ** Montandon 1897. Anterior margin of the pronotum as in **Cryphocricinæ.** Rostrum short and

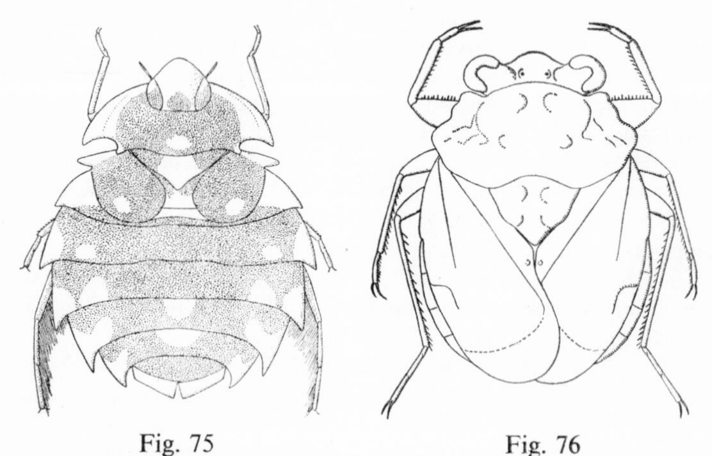

Fig. 75

Aphelochirus variegatus Kiritschenko 1925 (Naucoridæ-Aphelocheirinæ).

Fig. 76

Gelastocoris buto Herrich-Schaeffer 1837 (Gelastocoridæ-Gelastocorinæ).

articulated to the posterior portion of a deep excavation under the head and moderately widely spaced from the anterior margin of the vertex. Labrum not visible. Head convexly and laminately produced in front of eyes. **Potamocorinæ** Usinger 1941 very small, primitive Naucoridæ, doubtless allied to the Helotrephidæ. The head is not deeply retracted into the prothorax and the rostrum composed of three segments is slender. The antennæ are slender and composed of four segments. The gula is long and not carinate. The posterior margin of the pronotum is medially excavate. The anterior tarsi have one segment and the remaining tarsi two segments. All are armed with two claws. **Aphelocheirinæ** Fieber (1860) with the legs somewhat modified for raptorial purposes, the antennæ fairly long and slender, the head triangular and the rostrum very long, slender and extending beyond the median coxæ. In the male the genital segments are asymmetrical. The anterior tarsi have three segments, the basal segment very small and obscure.

The distribution of the Naucoridæ is world-wide.

References

Kiritshenko 1952; Larsen 1927; Ohm 1956; Thorpe and Crisp 1947; Usinger 1946c.

GELASTOCORIDÆ Kirkaldy 1897, *Entomologist*, **30,** 258 (Fig. 76)

Species belonging to this family are usually considered to be semi-aquatic since not only water but wet mud and vegetable debris have been noted as habitats.

The characters of representatives of this family are the somewhat batrachian-like appearance and manner, prominent eyes, raptorial anterior legs (but not strictly of the true raptorial type), antennæ with four segments concealed below the eyes. Ocelli present. Rostrum short. The male genital segments are asymmetrical. The dorsal surface of the thorax and hemelytra is often rugose.

Some species occur in stagnant water. *Mononyx nepæformis* (Fabricius) 1794 (**Mononychinæ**) has been discovered among vegetable debris and has also been observed climbing low vegetation presumably in order to find food which appears to be small soft-bodied insects including possibly termites. This species has been seen to feign death when disturbed. *Mononyx montandoni* Melin 1930 has been found in the wet mud of rice seedbeds.

The life history of *Gelastocoris oculatus* (Fabricius) 1798 (**Gelastocorinæ**) in the United States of America has been described by Hungerford. This species is found on muddy banks of streams or on sandy river beaches. It has the habit of burrowing in the sand or mud, probably for the purpose of avoiding being washed away in times of flood. Any kind of insect of suitable size appears to be acceptable to them as prey.

The females oviposit either on or in sand or mud and produce about two hundred ova during the season. The ovum is broadly oval with the chorion granular and reticulate (Fig. 46, 27). On eclosion it splits lengthwise for part of its length. It is not known whether an egg-burster is present.

There are two subfamilies: **Gelastocorinæ** Champion 1901 which have the anterior tarsi provided with two claws, the hemelytra distinctly divided into corium and membrane, the pronotum with the lateral margins strongly sinuate, the widest part at the middle, the anterior pronotal margin also sinuate, the head not sunk beneath the anterior pronotal margin, the eyes extremely prominent and raised, pedunculate; **Mononychinæ** Fieber 1851, in which the anterior tarsi are armed with a single claw, the pronotum has the lateral margins sub-parallel, feebly convex and the anterior margin behind the vertex straight. The head is partly sunk beneath the anterior pronotal margin and the eyes are prominent but not pedunculate.

References Hungerford 1922; Kevan 1942.

OCHTERIDÆ Kirkaldy 1906, *Trans. Amer. Ent. Soc.* **32,** 149 (Fig. 77)

Insects which are apparently semi-aquatic and which have a short, stumpy ovate body, the dorsal surface of which is pubescent, a long rostrum extending at least to the posterior coxæ, the antennæ visible from above, the anterior legs similar to median and posterior legs, cursorial, the head with the vertex a little wider than the diameter of an eye, ocelli present, the scutellum flattened and tumid and the hemelytral membrane with large, pentagonal cells in two series.

Judging by the mouthparts and the anterior legs which are of a simple type, it would seem that the Ochteridæ are phytophagous, but in fact they are predaceous.

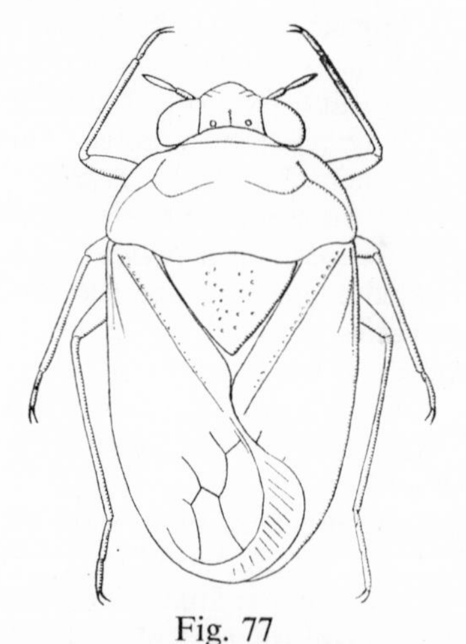

Fig. 77
Ochterus marginatus Latreille 1807 (Ochteridæ).

Observations on *Ochterus marginatus* Latreille 1807 show that the adults live on the shady shores of ponds and streams. They do not enter the water except by accident; but on the other hand, the neanides are amphibious and often submerge themselves. The neanides cover their dorsum with sandy granules and when about to moult they construct small cells in the sand into which they retire until ecdysis is complete.

When the neanides are submerged they maintain their bodies just under the surface film of the water. Before submersion they collect air on the ventral surface of the abdomen and when a fresh supply is required they rapidly turn over on their backs to expose the abdomen.

When about to copulate the male grasps the female with the second pair of legs, the anterior pair being fixed beneath the body and the abdomen somewhat to the left side. The reason for this fixed position is that the male genitalia are asymmetrical. According to observations, copulation is frequent and pairs may remain *in coitu* for as long as seven hours.

The genus *Ochterus* Latreille 1807 is cosmopolitan and *Megochterus* Jaczewski 1933 is Australian.

References Takahashi 1921.

References

Abalos J. W. and Wygodzinsky P. 1951, *Las Triatominæ Argentinas* (*Reduviidæ-Hemiptera*). Monog. 2, Instituto de Medicina Regional, Tucumàn.

Abbot J. F. 1912, *Rhamphocorixa balanodis* Abbott. *Amer. Nat.* **46**, 553-5.

Amyot C. J. B. and Serville A. 1843, *Hist. nat. Hémiptères* (suites de Buffon).

Ashlock P. D. and Lattin J. D. 1963, 'Stridulatory mechanisms in the Lygæidæ, with a new American genus of Orsillinæ (Hemiptera, Heteroptera)'. *Ann. ent. Soc. Amer.* **56**, 693-703.

Bacot A. 1921, 'Bionomics of *Cimex hirundinis*'. *Proc. Roy. ent. Soc. Lond.* 2.

Bailey N. S. 1951, 'The Tingoidea of New England and their biology'. *Ent. Americana*, **31** (new series), 1-140.

Balduf W. V. 1939, 'Food Habits of *Phymata pennsylvanica americana* Melin. (Hemiptera)'. *Canad. Ent.* **71**, 66-74.

Balduf W. V. 1941, 'Life History of *Phymata pennsylvanica americana* Melin' *Ann. ent. Soc. Amer.* **34**, 204-14.

Banks C. J. 1938, 'On the occurrence of nematodes in *Corixa geoffroi* Leach (Hem. Corixidæ)'. *J. Soc. Brit. Ent.* 217-9.

Baranowski R. M. 1958, 'Notes on the biology of the Royal Palm Bug *Xylastodoris luteolus* Barber (Hemiptera-Thaumastocoridæ)'. *Ann. ent. Soc. Amer.* **51**, 547-51.

Barber G. W. 1923, 'Notes on a New Zealand Aradid (*Aradus* 4-*lineatus*)'. *Psyche, Camb. Mass.* **30**, 120-2.

Baudoin R. 1939, 'Sur l'Habitat d'*Aepophilus bonnairei* Signoret en deux stations des côtes de France'. *Bull. Soc. Zool. France*, **64**, 18-20.

Beament J. W. L. 1946, 'The formation and structure of the chorion of the egg in an Hemipteran, *Rhodnius prolixus*'. *Quart. J. Micr. Sci.* **87**, 393-439.

Beament J. W. L. 1947, 'The formation and structure of the micropylar complex in the egg-shell of *Rhodnius prolixus* Stahl (*sic*), Heteroptera-Reduviidæ'. *J. exp. Biol.* **23**, 213-33.

Behr L. 1852, 'Über die Blütengalle des *Teucrium chamædrys*'. *L. Deutsch. Bot. Ges.* **65**, 326-30.

Bergroth E. 1886, 'Zur Kenntnis der Aradiden'. *Verh. zool.-Bot. Ges. Wien* **36**, 53-60.

Bergroth E. 1892, 'Aradidi dell'isola di Engano raccolti dal Dott. Elio Modigliani'. *Ann. Mus. Stor. nat. Genova* (2), **12**, 806-8.

Bergroth E. 1899, 'Note on the genus *Aepophilus* Sign'. *Ent. month. Mag.* **35**, 282.

Bergroth E. 1903. 'Neue myrmecophile Hemipteren'. *Wien ent. Ztg.* **22**, 255-6.

Bergroth E. 1914, 'On an Hemipterous insect from an Australian opossum's nest'. *Trans. roy. Soc. S. Aust.* **38**, 53-7.

Berlese A. 1914, *Gli Insetti* **2**, fasc. 7-8, 220.

Blanchard R. 1902, 'Sur la piqure de quelques Hémiptères'. *Arch. Parasit. Paris*, **5**, 1, 139-48.

Blöte H. C. 1945, 'On the systematic position of *Scotomedes* (Heteroptera-Nabidæ)'. *Zoöl. Meded.* **24**, 321-4.

Bodenheimer F. S. 1951, *Insects as human food.*

Bolivar I. 1894, 'Observations sur la *Phyllomorpha laciniata* de Villiers. *Feuille. jeun. Nat.* **24**, 43.

Brien P. 1930, 'Notes sur *Phloea paradoxa* Burm. in *Mission Biologique belge au Brésil* (1922-1923), **2**, 203-12.

Brindley M. D. H. 1930, 'On the metasternal scent-glands of certain Heteroptera'. *Trans. ent. Soc. Lond.* **78**, 199-207.

Brindley M. D. H. 1934, 'A note on the eggs and breeding habits of *Salda littoralis*'. *Proc. Roy. ent. Soc. Lond.* **9**, 10-11.

Brown E. S. 1958a, 'Revision of the genus *Amblypelta* Stål (Hemiptera-Coreidæ)', *Bull. ent. Res.* **49**, 509-41.

Brown, E. S. 1957b, Injury to Cacao by *Amblypelta* Stål (Hemiptera-Coreidæ with a summary of food-plants of species of this genus). *Bull ent Res.* **49; 3**, 343-554.

Brown E. S. 1959a, 'Immature Nutfall of Coconuts in the Solomon Islands. I. Distribution of Nutfall in Relation to that of *Amblypelta* and of Certain Species of Ants'. *Bull. ent. Res.* **50,** 1, 97-133.

Brown E. S. 1959b, 'Immature Nutfall of Coconuts in the Solomon Islands. II. Changes in Ant Population and the Relation to Vegetation'. *Bull. ent. Res.* **50,** 3, 523-58.

Brown E. S. 1959c, 'Immature Nutfall of Coconuts in the Solomon Islands. III. Notes on the Life-History and Biology of *Amblypelta*'. *Bull. ent. Res.* **50,** 3, 559-66.

Brown E. S. 1959d, '*Salduncula,* an intertidal Saldid in Madagascar (Hemiptera)'. *Le Naturaliste Malgache,* **11, 1-2,** 73-6.

Brues Charles T., Melander A. L., Carpenter Frank M. 1954, 'Classification of Insects'. *Bull. Mus. comp. Zool. Harv.* **108.**

Brumpt E. 1912, 'Le trypanosome cruzi évolué chez *Conorhinus megistus, Cimex lectularius, Cimex boueti* et *Ornithodorus moubata,* cycle évolutif de ce parasite'. *Bull. Soc. Path. exot.* **5,** 360.

Brumpt E. 1914a, 'Importance du cannibalisme et de la coprophagie chez les Réduvidés hématophages (*Rhodnius, Triatoma*) pour la conservation des trypanosomes pathogènes en dehors de l'hôte vertèbré. *Bull. Soc. Path. exot.* **7,** 702-5.

Brumpt E. 1914b, 'Le Xénodiagnostic. Application diagnostic de quelques infections parasitaires et en particulier à la trypanosome de Chagas'. *Bull. Soc. Path. exot.* **7,** 706.

Buchanan-White F. 1877, '*Laccometopus clavicornis* L. and its relation to *Teucrium chamædrys*'. *Ent. mon. Mag.* **13,** 283.

Bueno I. R. de la Torre 1906, 'Life Histories of North American Water Bugs'. *Canad. Ent.* **38,** 189-97, 202-52.

Buchanan-White 1883, 'Report on the pelagic Hemiptera procured during the voyage of H.M.S. *Challenger* in the years 1873-1876'. *Zool. Chall. Exped.* 19.

Butler E. A. 1923, *A Biology of the British Hemiptera.*

Buxton P. A. 1930, 'The Biology of a blood-sucking bug *Rhodnius prolixus*'. *Trans. ent Soc. Lond.* **78,** pt. 2, 227-36.

Carayon J. 1949a, 'L'oothèque des Hémiptères Plataspidés de L'Afrique tropicale'. *Bull. Soc. ent. Fr.* **54.** 66

Carayon J. 1949b, 'Observation sur la Biologie des Hémiptères Microphysidés' *Bull. Mus. Paris,* 2e serie **21,** 710-16.

Carayon J. 1950a, 'Observation sur l'Accouplement, la Ponte et l'Éclosion chez des Hémiptères Henicocephalidés de l'Afrique tropicale'. *Bull. Mus. Paris,* 2e, serie **22,** 6.

Carayon J. 1950b, 'Caractères anatomiques et position systèmatique des Hémiptères Nabidés (note preliminaire)'. *Bull. Mus. Paris,* 2e serie **22,** 95-101.

Carayon J. 1951, 'Écologie et Régime alimentaire d'Hémiptères Henicocephalidés'. *Bull. Soc. ent. France,* 39-44.

Carayon J. 1952, 'Existence chez certains Hémiptères Anthocoridés d'un organe analogue à l'Organe de Ribaga'. *Bull. Mus. Hist. nat.* **24** (2), 89-97.

Carayon J. 1953a, 'Organe de Ribaga et Fécondation chez un Hémiptère Cimicidé du Cambodge, *Aphraniola orientalis* Ferris and Usinger'. *Rev. Fr. Ent.* **20,** 139-45.

Carayon J. 1953b, 'Organe de Ribaga et fécondation hémocœlienne chez les *xylocoris* du groupe *galactinus* (Hémipt-Anthocoridæ)'. *C.R. Acad. Sci. France,* **236,** 1099-101.

Carayon J. 1953c, 'Existence d'un double orifice génital et d'un tissu conducteur des spermatozoides chez les Anthocorinæ (Hemiptera-Anthocoridæ)'. *C. R. Acad. Sci. France* **236,** 1206-8.

Carayon J. 1954a, 'Fécondation hemocœlienne chez un Hémiptère Cimicidé dépourvu d'organe de Ribaga. *C. R. Acad. Sci. France*, **239**, 1542-4.

Carayon J. 1954b, 'Quelques Hémiptères Nabidés du Congo Belge'. *Ann. Mus. Congo, Zool.* **1**, 320-5.

Carayon J. 1954c, 'Organes assumant la fonction de la spermathéque chez divers Hétéroptères'. *Bull. Soc. Zool. France*, LXXIX, 189-97.

Carayon J. 1956, 'Anthocoridæ, Scolopini nouveaux d'Afrique tropicale (Hémipt-Hetéropt.)'. *Bull. Mus. Paris*, **28**, 2e serie, 102-10.

Carayon J. 1957, 'Introduction à l'Etude des Anthocoridæ Omphalophores (Hemiptera-Heteroptera)'. *Ann. Soc. ent. Fr.* **126**, 159-96.

Carayon J. 1960, '*Stethoconus frappai* n. sp. Miridé prédateur du Tingidé du Cafeier, *Dulinius unicolor* (Sign.) à Madagascar'. *Journ. d'Agric. Trop. et de Botanique appliqué*, **7**, 110-20.

Carayon J. 1960, 'La Viviparité chez les Héteroptères'. Verh. XI, *Int. Kong. fur Ent. Wien*, **1**, 711-4.

Carayon J. 1962, 'Observations sur l'appareil odorifique des Hétéroptères, particulierement celui des Tingidæ, Vianaididæ et Piesmatidæ'. *Cahier des Naturalistes parisiens*, **1**, 1-16.

Carayon J., Usinger R. L. and Wygodzinsky P. 1958, 'Notes on the higher classification of the Reduviidæ with the description of a new tribe of the Phymátidæ'. *Rev. Zool. Bot. Afr.* **57**, 270.

Carvalho J. C. M. 1951, 'New Genera and Species of Isometopidæ in the Collection of the British Museum of Natural History (Hemiptera)'. *Ann. Acad. bras. Sci.* **23**, 390.

Carvalho J. C. M. 1952, 'On the major Classification of the Miridæ (Hemiptera) with keys to the sub-families and tribes and a Catalogue of the World Genera'. *Ann. Acad. bras. Sci.* **24**, 31-111.

Carvalho J. C. M. and China W. E. 1951, 'Neotropical Miridæ XLIII, A Remarkable ant-like genus from Uruguay, with notes on a new tribe of Mirinæ'. *Entomologist* **84**, 265-6.

Chen S. H. and Young B. 1943,, Further remarks on the carriage of eggs on the elytra of the males of *Sphærodema rusticum*'. *Sinensis* **14**, 49-53.

Champion G. C. 1898, *Biologia Centrali Americana, Hem.-Het.* **2**, 231.

China W. E. 1926, 'The egg of *Ploariola culiciformis* de Geer' *Ent. mon. Mag.* **62**, 265-6.

China W. E. 1927, 'Notes on the biology of *Aepophilus bonnairei* (Hemiptera-Heteroptera)'. *Ent. mon. Mag.* **63**, 238-41.

China W. E. 1930, 'The Origin of the British Heteropterous Fauna'. *Peuplement des Isles Britanniques*, Paris, 1930.

China W. E. 1931, 'Morphological Parallelism in the Structure of the labium in the Hemipterous genera *Coptosomoides* gen. nov. and *Bozius* Distant (Fam. Plataspidæ) in connection with mycetophagous habits'. *Ann. Mag. nat. Hist.* (10), 7, 281.

China W. E. 1933, 'A New Family of the Hemiptera-Heteroptera with notes on the phylogeny of the suborder'. *Ann. Mag. nat. Hist.* (10), 12, 180-96.

China W. E. 1943, 'The Generic Names of British Insects' **8**. Hemiptera-Heteroptera, 217-325.

China W. E. 1945, 'A completely blind bug of the family Lygæidæ (Hemiptera Heteroptera'. *Proc. Roy. ent. Soc. Lond.* **14**, 126-8.

China W. E. 1946, 'New Cryptostemmatidæ (Hemiptera) from Trinidad, British West Indies'. *Proc. Roy. ent. Soc. Lond.* Ser. B **15**, 148-54.

China W. E. 1953a, 'A new subfamily of Microphysidæ (Hemiptera-Heteroptera)'. *Ann. Mag. nat. Hist.* (12), **6**, 67.

China W. E. 1953b, 'Two new species of the genus *Cyrtopeltis* (Hemiptera) associated with sundews in Western Australia'. *W. Aust. Nat.* **4**, 1, 1-12.

China W. E. 1954, 'Notes on the nomenclature of the Pyrrhocoridæ (Hemiptera-Heteroptera)'. *Ent. mon. Mag.* **90**, 188.

China W. E. 1955a, 'A Reconsideration of the Classification of the Joppeicidæ with Notes on the Phylogeny of the Heteroptera. *Ann. Mag. nat. Hist.* (12) 8, 257-67, (12), 8, 353-70.

China W. E. 1955b, 'The Evolution of the Waterbugs'. *Nat. Inst. Sci. India Bull.* VII, 91-103.

China W. E. 1955c, 'A New Genus and Species representing a new subfamily of Plataspidæ with notes on the Aphylidæ (Hemiptera-Heteroptera)'. *Ann. Mag. nat. Hist.* (12) 8, 204-210.

China W. E. 1957, 'The Marine Hemiptera of the Monte Bello Islands with descriptions of some allied species'. *Journ. Linn. Soc. Zool.* 43, 291-352.

China W. E. and Carvalho J. C. M. 1951, 'A New Ant-like Mirid from Western Australia (Hemiptera, Miridæ)'. *Ann. Mag. nat. Hist.* (12), 4, 221.

China W. E. and Miller N. C. E. 1955, 'Check List of Family and Subfamily names of the Hemiptera-Heteroptera'. *Ann. Mag. nat. Hist.* (12), 8, 257-67.

China W. E. and Miller N. C. E. 1959, 'Check List and Keys to the Families and Subfamilies of the Hemiptera-Heteroptera. *Bull. Brit. Mus.* (*N. H.*) *Entomology* **8,** 1, 1-45.

China W. E. and Usinger R. L. 1949, 'A New Genus of Tribelocephalinæ from Fernando Poo (Hemiptera-Reduviidæ)'. *Ann. Mus. Stor. nat. Genova* **64,** 43-7.

China W. E. and Slater James A. 1956, 'A New subfamily of Urostylidæ from Borneo (Hemiptera-Heteroptera)'. *Pacific Science* **10,** 410-4.

China W. E. and Myers J. G. 1929, 'A Reconsideration of the Classification of the Cimicoid families'. *Ann. Mag. nat. Hist.* (10), 3, 105-18.

Cobben R. H. 1957, 'Beitrag zur Kenntnis der Uferwanzen (Hem. Het. Fam. Saldidæ)'. *Ent. Ber.* **21,** 96-107.

Cobben R. H. 1968, 'Evolutionary Trends in Heteroptera'. *Wageningen.*

Conradi Albert F. 1904, 'Variations in the protective value of odoriferous secretions of some Heteroptera'. *Science* (n.s.) **19,** 393-4.

Cook A. J. 1897, 'Another Bee Enemy'. *Canad. Ent.* **11,** 17-20.

Coons G. H., Kotila J. E. and Stuart D. 1957, 'Savoy, a virus disease of Beet transmitted by *Piesma cinerea*'. *Phytopathology* **27,** 125.

Corbett G. H. and Miller N. C. E. 1933, 'A List of Insects with their Parasites and Predators in Malaya'. *Sci. Ser. Dept. Agric. S.S.* and *F.M.S.* 13.

Corby H. D. L 1947, '*Aphanus* (Hem-Lygæidæ) in Stored Groundnuts'. *Bull. ent. Res.* **37,** 609-17.

Costa Lima A. da, Campos C. A., Hathaway C. R. 1951, 'Estudo dos Apiomeros'. *Mem. Inst. Osw. Cruz* **49,** 273-442.

Cott H. B. 1934, 'The Zoological Society's Expedition to the Zambesi 1947, No. 5. On a collection of Lizards from Portuguese East Africa with descriptions of new species of *Zonarus*, *Monopeltis* and *Chirindea*'. *Proc. zool. Soc. Lond.* 145-73.

Cuthbertson A. 1934, 'Note on the swarming of Pentatomid Bugs'. *Nada* Ann. Native Affairs Dept. S. Rhodesia, 38.

Dahms R. G. and Kagan M. 1938, 'Egg parasite of the Cinch Bug'. *J. Econ. Ent.* **31,** 779-80.

Davis Ch. C. 1961, 'A Study of the Hatching Process in Aquatic Vertebrates II, Hatching in *Ranatra fusca* P. Beauvois (Hemiptera, Nepidæ)'. *Trans. Amer. Microsc. Soc.* **80,** 230-4.

Davis Ch. C. 1964, 'A Study of the Hatching Process in Aquatic Vertebrates VII, Observations on Hatching in *Notonecta melæna* Kirkaldy (Hemiptera Notonectidæ) and on *Ranatra absona* D. and Dec. (Hemiptera Nepidæ), The Hatching Process of *Amicola hydrobioides* (Ancey), (Prosobranchia Hydrobiidæ).' *Hydrobiologia* **23,** 253-66.

Davis Ch. C. 1965, 'A Study of the Hatching Process in Aquatic Vertebrates XII, The Eclosion Process in *Trichocorixa naias* (Kirkaldy (Heteroptera Corixidæ)'. *Trans. Amer. Microsc. Soc.* **84,** 6065.

Delamare Deboutteville C. and Paulian R. 1952, 'Recherches sur la Faune des Nids et des Terriers en Basse Côte d'Ivoire'. *Encyc. Biogeog. écol. Paris* **8,** 116 pp.

Deventer W. van 1906, 'De Dierlijke Vijanden van het Suikerreit en hunne Parasieten. Handboek te Dienst van de Suikerreit-Cultuur in de Rietsuiker-Fabricage op Java'. 2 deel.

Diakonoff A. 1841, *Arch. Suikerind. Ned.-Ind.* **2,** 205-13.

Dispons P. 1951, 'Phototropisme positif chez *Oncocephalus squalidus* Rossi'. *L'Entomologiste* **7,** 64.

Dispons P. 1955a, 'Les Réduvidés de l'Afrique Nord-Occidentale; Biologie et Biogeographie'. *Mem. Mus. Hist. nat. Paris* (A), **10,** 93-240.

Dispons P. 1955b, 'Observations sur la ponte de *Rhinocoris erythropus* (L.)'. *L.Entomologiste, Paris,* **11,** 14-7.

Distant W. L. 1903, *Fasc. Malay.* 2.

Distant W. L. 1904, *The Fauna of British India* (*Rhynchota*), **2.**

Dodd F. P. 1904, 'Notes on the Maternal Instinct in Rhynchota'. *Trans. ent. Soc. Lond.* 483-6.

Douglas J. W. 1865, 'On the occurrence of *Systellonotus triguttatus* (a hemipterous insect) in company with *Formica fusca*'. *Ent. mon. Mag.* **2,** 30-1.

Douglas J. W. 1877, 'The Economy of *Laccometopus clavicornis* L.'. *Ent. mon. Mag.* **13,** 236-7.

Douglas J. W. and Scott J. 1865, *The British Hemiptera.*

Drake Carl J. 1925, 'An undescribed gall-making Hemiptera (Tingidæ) from Africa'. *Amer. Mus. Nov.* **158,** 2.

Drake Carl J. 1956, 'First Record of the family Leptopodidæ (Hemipt.) from Australia'. *Mem. Queensland Mus.* **13,** 2, 146-7.

Drake Carl J. and Davis Norman T. 1958, 'The Morphology and Systematics of the Piesmatidæ (Hemiptera) with Keys to the World Genera and American species'. *Ann. ent. Soc. Amer.* **51,** 6, 567-81.

Drake Carl J. and Davis Norman T. 1960, 'The Morphology, Phylogeny and Higher Classification of the Family Tingidæ, including the descripion of a new genus and species of the subfamily Vianaidinæ (Hemiptera; Heteroptera)'. *Ent. Americana,* **39,** 1-100.

Drake Carl J. and Slater James A. 1957, 'The Phylogeny and Systematics of the family Thaumastocoridæ (Hemiptera-Heteroptera)'. *Ann. ent. Soc. Amer.* **50,** 4, 353-70.

Dufour L. 1833, 'Récherches anatomiques et physiologiques sur les Hémiptères, etc.' *Mem. des Savants étrang. à l'Academie des Sciences* **4,** 129-462.

Dufour L. 1834, 'Observations sur le genre *Prostemma*'. *Ann. ent. Soc. Fr.* **3,** 350.

Edwards J. S. 1962, 'Observations on the development and predatory habits of two Reduviid Heteroptera, *Rhinocoris carmelita* Stål and *Platymeris rhadamanthus* Gerst'. *Proc. Roy. ent. Soc. Lond.* **37,** 89-98.

Eidmann H. 1924, 'Untersuchungen über den Mekanismus der Hautung bei den Insekten'. *Arch. f. Mikrosk. Anatomie und Entwicklungsmekanik. Berlin,* **102,** 276-90.

Ekblom Tore 1926, 'Morphological and Biological Studies of the Swedish Families of Hemiptera-Heteroptera'. *Zool. Bidr. Uppsala* 10.

Elson J. A. 1937, 'A Comparative Study of Hemiptera'. *Ann. ent. Soc. Amer.* **30,** 579-97.

Esaki T. 1924, 'On the curious halophilous waterstrider *Halovelia maritima* Bergroth (Hemiptera-Gerridæ)'. *Bull. Brooklyn ent. Soc.* **19,** 29-34.

Esaki T. 1930, 'Marine waterstriders of the Corean Coast (Hemiptera-Gerridæ)'. *Ent. mon. Mag.* **65,** 158-61.

Esaki T. 1931, 'Eine neue Gallenbildende Tingiditen-Art'. *Bull. Sc. Fac. Tak. Kyusu Imp. Univ.* **4** (3), 244-53.

Esaki T. 1947, 'Notes on *Hermatobates haddonii* Carpenter (Hemiptera-Gerridæ)'. *Mushi* **18,** 7, 49-51.

Esaki T. and Matsuda R. 1951, 'Hemiptera Micronesica, 3, Dysodiidæ'. *Mushi* **22,** 73-86.

Evans J. W. 1948, *Insect Pests and their Control,* Tasmanian Department of Agriculture.

Eyles A. C. 1963, 'Descriptions of the immature stages of five Rhyparochrominæ (Heteroptera-Lygæidæ)'. *Trans. Soc. Brit. Ent.* **15,** 277-94.

Ferris G. F. and Usinger R. L. 1939, 'The Family Polyctenidæ (Hemiptera-Heteroptera)'. *Microentomology* **4,** 1, 1-50.

Fernando C. H. and Leong C. Y. 1963, 'Miscellaneous Notes on the Biology of Malayan Corixidæ (Hem. Heteroptera) and on a study of the life history of two species, *Micronecta quadristrigata* Bredd., and *Agraptocorixa hyalinipennis* (F.)'. *Ann. Mag. nat. Hist.* **13,** 545-558.

Fox-Wilson G. 1925, 'The egg of the tarnished plant bug *Lygus pratensis* Linn'. *Ent. mon. Mag.* **61,** 19.

Frauenfeld G. R. 1853, 'Über die Pflanzenauswuchse von *Teucrium montanum* und *Laccometopus* (*Cimex*) *teucrii*'. *Verh. Zool.-Bot. Verein, Wien,* **3,** 157-161.

Frost W. E. and Macan T. T. 1948, 'Corixidæ as food of Fish'. *J. Anim. Ecol.,* 174-79.

Gadeau de Kerville H. 1902, 'L'Accouplement des Hémiptères'. *Bull. Soc. ent. Fr.* **1920,** 67-71.

'Game Department of Uganda, Report', 1949. Quoted in *Oryx,* **1,** No. 2, 1951.

Gillett J. D. 1957, 'On the habits and life history of captive Emesine Bugs (Hemiptera-Reduviidæ)'. *Proc. Roy. ent. Soc. Lond.* **32,** 193-95.

Gillett J. D. and Wigglesworth V. B. 1932, 'The Climbing Organ of an Insect *Rhodnius prolixus* (Hemiptera-Reduviidæ)'. *Proc. Roy. Soc. B.* **111,** 365-75.

Girault A. A. 1906, 'Standards of the number of eggs laid by insects'. *Ent. News.* **17,** 6.

Grandi G. 1951, *Introd. Stud. Ent.* **1,** 337.

Griffith M. E. 1945, 'The environment, life history and structure of the water-boatman *Rhamphocorixa acuminata* (Uhler) (Hemiptera, Corixidæ). *Kansas Univ. Sci. bull.* **30,** 241-366.

Gross G. E. 1955, 'A Revision of the Flower Bugs (Heteroptera, Anthocoridæ) of the Australian and Adjacent Pacific Regions'. Pt. 2. *Records S. Aust. Mus.* **11,** 4, 407-22.

Gross J. 1901, 'Untersuchungen über das Ovarium der Hemipteren zugleich ein Betrag zur Amitosenfrage'. *Z. wiss. Zool.* **69,** 139-201.

Gupta A. P. 1961, 'A critical review of the studies on the so-called stink or repugnatorial glands of Heteroptera with further comments'. *Canad. Ent.* **93,** 482-86.

Hagen H. R. 1931, 'The Embryogeny of the Polyctenid *Hesperoctenes fumarius* Westwood with reference to viviparity in insects'. *J. Morph.* **51,** 1-92.

Halaszfy Eva 1959, 'Bermerkungen über einige stadien der larven von Pentatomoidea in der Zoologischen Abteilung des Ungarischen National-Museums'. *Acta Ent.* **32,** 526, 545-66.

Hale H. M. 1924, 'Notes on eggs, habits and migration of some Australian aquatic bugs (Corixidæ and Notonectidæ)'. *S. Aust. Nat.* **5,** 133-35.

Handlirsch A. 1897, 'Über *Phimophorus spissicornis* Bergr. Ein Hemipterologische Beitrag'. *Verhandl. Zool. Bot. Gesell. Wien,* **47,** 408-10, 2 figs.

Handlirsch 1900a, 'Zur Kenntnis der Stridulationsorgane bei den Rhynchoten'. *Ann. Nat. Hofm. Wien,* **15,** 127-41.

Handlirsch A. 1900b, 'Neue Beitrage zur Kenntnis der Stridulationsorgane bei den Rhynchoten'. *Verh. zool.-bot. Ges. Wien* **50,** 555-60.

Handlirsch A. 1908, *Die fossilen Insecten.*

Harris H. M. 1928, 'A monographic study of the Hemipterous family Nabidæ as it occurs in North America'. *Ent. amer.* **9,** 1-98.

Hase A. 1917, 'Die Bettwanze (*Cimex lectularius* L.) ihr leben and ihre Bekampfung'. *Monog. zur angew. Ent. beihefte zur Zeitschr. f. angew. Ent.* **4.**

Hase, A. 1933, 'Uber Starrezustande bei Blutsaugenden Insekten insbesondere bei Wanzen II. Mitteilung betr. *Panstrongylus* (*Triatoma*) *geniculatus* Pinto 1931 (Hem.)'. *Zeitschr. fur Parisit.* **5,** 3-4, 708-23.

Hase A. 1941, 'Weitere Beobachtungen uber Maskierung bei Parasitaren Wanzen (Cimicidæ und Triatomidæ)'. *Zeitschr. fur Parasit.* **12**, 3, 388-403.

Heidemann O. 1911, 'Some remarks on the eggs of North American species of Hemiptera-Heteroptera'. *Proc. ent. Soc. Wash.* **13**, 128-40.

Hemming F. 1953, 'Copenhagen Decisions on Zool. Nomenclature'. *Intern. Trust Nomencl. London.*

Henneguy L. F. 1904, *Les Insectes.*

Herrer A., Lent H. and Wygodzinsky P. 1954, 'Contribucion al conocimiento del genero *Belminus* Stål 1959 (Triatominæ, Reduviidæ, Hemiptera)'. *Anales del Instituto de Medecina Regional de la Univ. Nac. de Tucumàn*, **4**, 85-106.

Herring Jon L. 1949, 'A New species of *Rheumatobates* from Florida (Hemiptera-Gerridæ)'. *Florida Ent.* **32**, 160-5.

Hesse A. J. 1940, 'A New Species of *Triphleps* (Hemiptera-Heteroptera-Anthocoridæ) predaceous on the citrus thrips (*Scirtothrips aurantii* Faure) in the Transvaal'. *J. ent. Soc. S. Africa*, **3**, 61-71.

Hesse A. J. 1947, 'A remarkable new dimorphic Isometopid and two other species of Hemiptera predaceous upon the red scale of citrus'. *J. ent. Soc. S. Africa*, **10**, 31-45.

Heymons R. 1906, 'Über einen Apparat zum offnen der Eischalen bei den Pentatomiden'. *Z. wiss. Insekt. Biol.* **2**, 73.

Heymons R. 1926, 'Über Eischalensprenger und den Vorgang des Schlüpfens aus der Eischale bei den Insekten'. *Biol. Zentralblatt* **46**, 51-63.

Hibrauoi M. 1930, 'Contribution à l'étude biologique et systématique de *Eurygaster integriceps* Put. en Syrie'. *Rev. Path. veg. Ent. Agric.* **17**, 97-160.

Hill A. R. 1957, 'The Biology of *Anthocoris nemorum* (L.) in Scotland (Hemiptera-Anthocoridæ)'. *Trans. Roy. ent. Soc. Lond.* **109**, (13), 379-94.

Hinton H. E. 1962, 'The structure of the shell and respiratory system of the eggs of *Helopeltis* and related genera (Hemiptera, Miridæ)'. *Proc. Zool. Soc. Lond.* **139**, 483-8.

Horváth G. 1894, 'Sur la stridulation de *Spathocera laticornis* Schill. *Feuille. jeun. Nat.* **24**, 90.

Horváth G. 1906, 'A new gall-inhabiting bug from Bengal'. *Ent. mon. Mag.* **42**, 33-4.

Horváth G. 1911, 'Nomenclature des familles des Hémiptères'. *Ann. Mus. Nat. Hung.* **9**, 1-34.

Horváth G. 1912, 'Sur les noms des familles et des sousfamilles du regne animale'. Verh. 8. *Int. Cong. Zool.* 1910.

Houard C. 1906, 'Sur les modifications histologiques apportés aux fleurs de *Teucrium chamædrys* et du *Teucrium montanum* par les larves des *Copium*'. *C.R. Acad. Sci. Paris*, **143**, 927-9.

Hungerford H. B. 1919, 'The Biology and Ecology of Aquatic and Semiaquatic Hemiptera'. *Kansas Univ. Sci. Bull.* **11.**

Hungerford H. B. 1948, 'The Corixidæ of the Western Hemisphere'. *Kansas Univ. Sci. Bull.* **32.**

Hungerford H. B. 1923, 'Notes on the Eggs of Corixidæ'. *Bull. Brooklyn ent. Soc.* **18**, 1, 14.

Hutchinson G. E. 1931, 'On the Occurrence of *Trichocorixa* Kirkaldy (Corixidæ, Hemiptera, Heteroptera) in salt water and its geographical significance'. *American Naturalist* **65**, 573-4.

Immel R. 1955, 'Zur Biologie und Physiologie von *Reduvius personatus* L. *Z. Morph. Okol. Tiere*, **44**, 163-95.

Imms A. D. 1934, *A General Textbook of Entomology* (3rd ed.).

Jacobson E. 1911, 'Biological Notes on the Hemipteron *Ptilocerus ochraceus*'. *Tijdschr. Ent.* **54**, 175-9.

Jaczewski T. 1961, 'Notatki z biologi wioslakow (Corixidæ-Heteroptera). Notes on the Biology of Corixidæ (Heteroptera)'. *Bull. ent. de Pologne*, **31**, 295-300.

Jeannel R. 1919, 'Voyage de Ch. Alluaud et R. Jeannel en Afrique orientale (1911-1912)'. *Insectes, Hémiptères*, 3.

Jeannel R. 1942, 'Les Hénicocephalidés. Monographie d'un groupe d'Hémiptères Hématophages'. *Ann. Soc. ent. Fr.* **110,** 273-368.

Johnson C. G. 1934, 'On the eggs of *Notostira erratica* L.' *Trans. Soc. Brit. Ent.* **1,** 1-32.

Johnson C. G. 1936, 'The Biology of *Leptobyrsa rhododendri* Horváth (Hemiptera, Tingidæ), the Rhododendron lacebug'. *Ann. Appl. Biol.* **23,** 324-68.

Jones P. A. and Coppel H. C. 1963, 'Immature Stages and Biology of *Apateticus cynicus* Say (Hemiptera-Pentatomidæ)' *Canad. Ent.* **95,** 770-9.

Jordan K. 1911, 'Polyctenidæ viviparous'. *Proc. ent. Soc. Lond.* lxiv.

Jordan 1913, 'On Viviparity in Polyctenidæ'. *Trans. 2nd Congress of Entomology, Oxford*, **2,** 342-50.

Jordan K. 1922, 'Notes on the distribution of the organ of Berlese in Clinocoridæ'. *Ectoparasites* **1,** 284-6.

Jordan K. H. C. 1932, 'Beitrag zur Kenntnis der Eier und Larven von Aradiden'. *Zool. Jahrb.* **63,** 3, 281-99.

Jordan K. H. C. 1932, 'Zur Kenntnis des Eier und der Larven von *Microvelia Schneideri*'. *Z. wiss. Insekt. Biol.* **127,** 18-22.

Jordan K. H. C. 1935, 'Beitrag zur Lebensweise der Wanzen auf feuchten boden (Heteroptera)'. *Stettin ent. Zeit.* **96,** 1-26.

Jordan K. H. C. 1937, 'Lebensweise und Entwicklung von *Micronecta minutissima* L. (Hem. Het.)'. *Ent. Jahrb.* **46,** 173-7.

Jordan K. H. C. 1940, 'Eine Bemerkungen über Cryptostemmatidæ (Hem. Het.).' *Ent. zeit.* **53,** 341-4.

Jordan K. H. C. 1959, 'Die Biologie von *Elasmucha grisea* (Heteroptera; Acanthosomidæ)'. *Beitr. Ent.* **8,** 385-97.

Jordan K. H. C. and Wendt A. 1938, 'Zur Biologie von *Salda littoralis* L. (Hem. Het.)'. *Stettin ent. Zeit.* **99,** 273-92.

Jourdan M. L. 1935, '*Clytiomyia helluo* F. parasite d'*Eurygaster austriaca* Schr. (Diptera-Tachinidæ)'. *Rev. franç. Ent.* **2,** 83-5.

Kalshoven L. G. E. 1950, *De Plagen van de Cultuurgewassen in Indonesie I.*

Kamenkova K. V. 1956, 'Parasites of Bugs of the Family Pentatomidæ in the Province of Krasnodar'. *Rev. d'Ent. U.R.S.S.* **35,** Pt. 2, 324-33.

Kellen W. R. 1959, 'Notes on the Biology of *Halovelia mariannarum* Usinger in Samoa (Veliidæ-Heteroptera)'. *Ann. ent. Soc. Amer.* **52,** 53-62.

Kellen W. R. 1960, 'A New species of *Omania* from Samoa with notes on its biology (Heteroptera; Saldidæ)'. *Ann. ent. Soc. Amer.* **53,** 494-9.

Kershaw J. C. W. 1910, 'On the Metamorphosis of two Coptosomine Hemiptera from Macao'. *Ann Soc. ent. Belg.* **54,** 69-73.

Kershaw J. C. W. and Kirkaldy G. W. 1908, 'On the Metamorphosis of two Hemiptera-Heteroptera from Southern China'. *Trans. Roy. ent. Soc. Lond.* Pt. 2, 59-62.

Kershaw J. C. W. and Kirkaldy G. W. 1909a, 'Biological Notes on Oriental Hemiptera No. 3'. *J. Bombay nat. Hist. Soc.* Aug. 15, 333-6.

Kershaw J. C. W. and Kirkaldy G. W. 1909b, 'Biological Notes on Oriental Hemiptera'. *J. Bombay nat. Hist. Soc.* Nov. 15, 571a-73.

Kevan D. Keith McE. 1942, 'Some Observations on *Mononyx nepæformis* (Fabricius), 1775. A Toad Bug (Mononychidæ, Hemip. Heteropt.'. *Proc. Roy. ent. Soc. Lond.* (A), **17,** 109-110.

Kiritschenko A. N. 1913, *Insecta Hemiptera. Faune de la Russie et des pays limitrophes.* 1-395.

Kiritschenko A. N. 1949, 'Nests of birds as biotope of the true Hemiptera'. *Ent. Obozr. Moscow* **30,** 239-41.

Kiritschenko A. N. 1952, 'New information on rheophilous true Hemiptera of the family Aphelocheiridæ of the continental waters of the U.S.S.R. New Information on the ecology of the Aphelocheiridæ (Hemiptera-Heteroptera) of the U.S.S.R.'. *Ent. Obozr. Moscow* **32,** 208-9.

Kirkaldy G. W. 1899, 'A Guide to the Study of British Waterbugs'. *Entomologist*, **29**, 3.

Kirkaldy G. W. 1900, 'Notes on some Sinhalese Rhynchota'. *Entomologist* **30**, 295.

Kirkaldy G. W. 1901, 'The Stridulating Organs of Waterbugs (Rhynchota) especially of Corixidæ'. *J. Quekett Micr. Cl.* **4**, 33-46.

Kirkaldy G. W. 1907, 'Biological Notes on the Hemiptera of the Hawaiian Islands'. No. 1, *Proc. Hawaii ent. Soc.* **1**, (4), 135-61.

Kirkaldy G. W. 1911, 'Some Remarks on the Reduviid Subfamily Holoptilinæ and on the species *Ptilocerus ochraceus* Montd'. *Tijdsch. Ent.* **54**, 170-4.

Kirkaldy G. W. 1906, 'List of the genera of Pagiopodous Hemiptera Heteroptera'. *Trans. Amer. ent. Soc.* **32**, 47-156.

Kirkpatrick T. W. 1923, 'The Egyptian Cotton Seed bug (*Oxycarenus hyalinipennis* Costa)'. *Bull. Minist. Agric. Egypt* 35.

Kirkpatrick T. W. 1935-36, '*Strepsiptera* attacking *Antestia*'. *Rep. E. Afr. Agric. Res. St.* **8**, 14-16.

Knowlton G. F. 1944. 'Pentatomidæ eaten by Utah Birds'. *J. Econ. Ent.* **37**, 118-19.

Knowlton G. F. and Nye W. P. 1946, 'Some Insect Food on the Sage Sparrow'. *J. Kansas ent. Soc.* **19**, 4, 139.

Kobayashi 1951, 'The developmental stages of some species of the Japanese Pentatomoidea (Hemiptera) IV', *Trans. Shikoku ent. Soc.* **2** (1), 7-16.

Kobayashi 1953, 'The developmental stages of some species of the Japanese Pentatomoidea (Hemiptera)'. *Sci. Rep. Matsuyama Agric. Coll.* **11**, 73-86.

Kobayashi 1954, 'The developmental stages of some species of the Japanese Pentatomoidea (Heteroptera), IV.' *Trans. Shikoku ent. Soc.* **4**, Pt .4, 79-82.

Kobayashi 1955, 'The developmental stages of some species of the Japanese Pentatomoidea (Hemiptera)'. *Trans. Shikoku ent. Soc.* **4**, 63-8.

Kormilev N. A. 1948, 'Una especia nueva de la familia Elasmodemidæ Leth. and Serv. (1896) de la Republica Argentina (Hemiptera-Heteroptera) (Reduvioidea)'. *Rev. ent. Soc. Arg.* **14**, 141-7.

Kormilev N. A. 1955, 'A new Myrmecophil family of Hemiptera from the delta of the Rio Parana, Argentina'. *Rev. Ecuat. Ent.* **2** (3-4), 465-77.

Krause G. 1939, 'Die Eitypen der Insekten'. *Biol. Zbb.* **59**, 495-536.

Kullenberg B. 1942, 'Die Eier der Swedischen Capsiden (Rhynchota), I.' *Ark. Zool.* **33**, No. 15, 1-15.

Kullenberg B. 1943, 'Die Eier der Swedischen Capsiden (Rhynchota) II.' *Ark. Zool.* **34a**, No. 15, 1-8.

Kullenberg B. 1944, 'Studien uber die Biologie der Capsiden'. *Zool. Bidr. Uppsala*, **23**, 1-522.

Kullenberg B. 1947, 'Uber Morphologie und Funktion der Kopulations-apparats der Capsiden und Nabiden'. *Zool. Bidrag. Uppsala*, **24**, 217-418.

Kunckel d'Herculais J. 1879, 'Observations sur les Mœurs et Metamorphoses du *Gymnosoma rotundatus* L.' *Ann. Soc. ent. Fr.* (5), **9**, 349-57.

Lanck David R. 1959, 'The Locomotion of *Lethocerus* (Hemiptera-Belostomatidæ'. *Ann. ent. Soc. Amer.* **52**, 93-9.

Larsen O. 1927, 'Über die Entwicklung und Biologie von *Aphelocheirus œstivalis*'. *Fabr. Ent. Tidskr.* **48**, 181-206.

Larsen O. 1930, 'Biologische Beobachtungen an Swedischen Notonecta-Arten'. *Ent. Tidskr.* **51**, 219-47.

Lattin J. D. 1958, 'A stridulatory mechanism in *Araphe cicindeloides* Walker'. *Pan. Pacific Ent.* **34**, 217.

Lebrun D. 1960, 'Recherches sur la biologie et l'éthologie de quelques Héteroptères aquatiques'. *Ann. Soc. ent. Fr.* **1929**, 179-99.

Lent H. 1939, 'Sobre o hematofagismo de *Clerada apicicornis* e outros artropodos; sua importancia na trasmissaõ da doença de Chagas'. *Mem. Inst. Osw. Cruz.* **34**, 5.

Lent H. and Jurberg J. 1966, 'Os estadios larvares de *Phlœophana longirostris* (Spinola 1837), (Hemiptera-Pentatomidæ)'. *Rev. Brasil Biol.* **26,** 1-4.

Leston D. 1952a, 'Notes on the Ethiopian Pentatomidæ 2. A Structure of unknown function in the Sphærocorini Stål (Hem. Het.)'. *Entomologist* **85,** 179-80.

Leston D. 1952b, '*Oncotylus viridiflavus* Goeze (Hem. Miridæ) and its food-plant Knapweed'. *Entomologist* **85,** 19.

Leston D. 1953, 'Phlœidæ Dallas; Systematics and Morphology with Remarks on the Phylogeny of the "Pentatomoidea" and upon the position of *Serbana* Distant'. *Rev. bras. Biol.* **13,** (2), 121-40.

Leston D. 1954a, 'Strigils and Stridulation in Pentatomoidea (Hem.), some new data and a review'. *Ent. mon. Mag.* **90,** 49-56.

Leston D. 1954b, 'The eggs of *Anthocoris gallarum-ulmi* (de G.) (Hem. Anthocoridæ, and *Monanthia humuli* (F.) (Hem. Tingidæ) with Notes on the eggs of Cimicoidea and Tingidoidea'. *Ent. mon. Mag.* **89,** 99-102.

Leston D. 1955, 'Notes on Ethiopian Pentatomoidea XVIII. The eggs of three Nigerian Shieldbugs with a tentative summary of egg form in Pentatomoidea'. *Ent. mon. Mag.* **91,** 33-6.

Leston D. 1957, 'Stridulatory Mechanism in terrestrial species of Hemiptera Heteroptera'. *Proc. zool. Soc. Lond.* **128,** 369-86.

Leston D., Pendergrast J. G., Southwood T. R. E. 1954. 'Classification of the Terrestrial Heteroptera (Geocoridæ)'. *Nature, Lond.* **1749.**

Leuckart R. 1835, 'Über die Mikropyle und den feinern Bau der Schalenhaut bei den Insekteneiren'. *Arch. Anat. Physiol., Lpz.* 90-264.

Livingstone D. 1962, 'On the biology and immature stages of a sap-sucker on *Zizyphus jujuba, Monosteira minitula* Mont. a species new to India (Hem. Tingidæ)'. *Agra Univ. Journ. of Res.* (*Science*), **11,** 117-29.

Lundbeck 1914, 'Some remarks on the eggs and egg-deposition of *Halobates*'. *Mindeskrift for Japetus Steenstrup*, **27,** 13 pp.

Malhota C. P. 1958, 'Bionomics of *Serinetha augur* Fabr. and its association with *Dysdercus cingulatus* Fabr. the red cotton-bug'. *Indian For.* **84,** 669:71.

Massee A. M. 1956, 'Hemiptera-Heteroptera associated with fruit and hops'. *Journ. Soc. Brit. Ent.* **5,** 179-86.

Maxwell-Lefroy H. 1909, *Indian Insect Life.*

Mckeown Keith C. 1934, 'Notes on the Food of Trout and Macquarie Perch in Australia'. *Rec. Aust. Mus.* **19,** 141-52.

Mckeown Keith C. 1943, 'The Foods of Birds from South-Western New South Wales'. *Rec. Aust. Mus.* **19,** 2.

Michalk O. 1934, 'Kannibalismus bei einen Pentatomide (Hem. Heteropt.), zugleich ein weiter Beitrag zur Technik der Nahrungsaufname der Wanzen'. *Ent. Z.* a.m. **48,** 51-5.

Michalk O. 1935, 'Zur Morphologie und Ablage der Eier bei den Heteropteren sowie über ein System der Eiablagetypen'. *Deutch. ent. Z.* 148-75.

Miller N. C. E. 1929a, '*Megymenum brevicorne* F. Pentatomidæ (Hem. Het.). A Minor Pest of Cucurbitaceæ and Passifloraceæ'. *Malay Agric. Journ.* **17,** 12, 421-36.

Miller N. C. E. 1929b, '*Physomerus grossipes* F. (Coreidæ, Hem. Het.). A Pest of Convolvulaceæ and Leguminosæ'. *Malay Agric. Journ.* **17,** 11, 403-20.

Miller N. C. E. 1931a, '*Geotomus pygmœus* Dallas (Heteroptera-Cydnidæ) attempting to suck human blood'. *Entomologist* **64,** 214.

Miller N. C. E. 1931b, 'The Bionomics of some Malayan Rhynchota (Hem. Het.) *Sci. Ser. Dep. Agric. S.S.* and *F.M.S.* 5.

Miller N. C. E. 1932a, 'Observations on *Melamphaus faber* F. (Hem. Pyrrhocoridæ) and descriptions of early stages'. *Bull. ent. Res.* **23,** 2, 195-201.

Miller N. C. E. 1932b, 'A Preliminary List of some Foodplants of some Malayan Insects'. *Bull. Dept. Agric. F.M.S.* 38.

Miller N. C. E. 1934, 'The Developmental Stages of Some Malayan Rhynchota'. *Journ. F.M.S. Mus.* **17,** 3, 502-25.

Miller N. C. E. 1937, 'A New Genus of Malayan Capsidæ (Rhynchota) from Areca Palm'. *Bull. ent. Res.* **28,** 4, 535-37.

Miller N. C. E. 1938a, 'A New Subfamily of Malaysian Dysodiidæ (Rhynchota)'. *Ann. Mag. nat. Hist.* (11), 1, 498-510.

Miller N. C. E. 1938b, 'Function of the *fossula spongiosa* or spongy furrow in Reduviidæ (Rhynchota)'. *Nature, Lond.* April 23.

Miller N. C. E. 1939, 'The *fossula spongiosa* in Reduviidæ'. *Nature, Lond.* Mar. 18, 477.

Miller N. C. E. 1941, 'Insects Associated with Cocoa (*Theobroma cacao*) in Malaya'. *Bull. ent. Res.* **32,** 1-15.

Miller N. C. E. 1942, 'On the Structure of the Legs in Reduviidæ (Rhynchota)'. *Proc. R. ent. Soc. Lond.* (A), **17,** 49-58.

Miller N. C. E. 1952, 'Three New Subfamilies of Reduviidæ (Hemiptera-Heteroptera), *Eos* **28,** 1, 88-90.

Miller N. C. E. 1953a, 'A Note on the Ova of Urostylidæ'. *Ent. month. Mag.* **89,** 137.

Miller N. C. E. 1953b, 'Notes on the Biology of the Reduviidæ of Southern Rhodesia'. *Trans. zool. Soc. Lond.* **27,** 541-656.

Miller N. C. E. 1953c, 'A New Subfamily and New Genera and Species of Australian Hemiptera-Heteroptera'. *Proc. Linn. Soc. N.S.W.* **77,** 233-40.

Miller N. C. E. 1954a, 'New Genera and Species of Reduviidæ from Indonesia and the Description of a New subfamily (Hem. Het.)'. *Tijdschr. v Ent.* **97,** 75-114.

Miller N. C. E. 1954b, 'A New subfamily and New Genera and Species of Malaysian Reduviidæ (Hem. Het.)'. *Idea* **10,** 1-8.

Miller N. C. E. 1955a, 'The Rostrum of *Centrocnemis* Signoret 1852 (Hemiptera-Heteroptera, Reduviidæ-Reduviinæ)'. *Nature, Lond.* **175,** 4458, 64.

Miller N. C. E. 1955b, 'The Synonymy of *Eupheno* Gistel (Hemiptera-Heteroptera, Reduviidæ) and the description of a New Subfamily" *Ann. Mag. nat. Hist.* (12), 8, 449-452.

Miller N. C. E. 1955c, 'The Metathoracic Gland Ostiole in *Eupheno* Gistel, *Cethera* Amyot and Serville, *Caprocethera* Breddin and *Carcinomma* Bergroth (Hemiptera-Heteroptera Reduviidæ)'. *Ann. Mag. nat. Hist.* (12), 8, 221-3.

Miller N. C. E. 1956a, 'Two New Species of Miridæ from the Agalega Islands'. *Mauritius Institute Bull.* **111,** 5, 317-20.

Miller N. C. E. 1956b, 'A New subfamily of Reduviidæ (Hemiptera-Heteroptera) from the Solomon Islands'. *Ann. Mag. nat. Hist.* (12) 9, 587-89.

Miller N. C. E. 1956c, 'Hemiptera Heteroptera, Reduviidæ'. *South African Animal Life* **111,** 434-92. Results of the Lund Univ. Exped. 1950-1951, Uppsala.

Miller N. C. E. 1956d, 'Centrocneminæ, A New Subfamily of the Reduviidæ (Hemiptera-Heteroptera)'. *Bull. Brit. Mus. Ent.* **4,** 6; 219-83.

Miller N. C. E. 1957a, 'New Genera and Species of Reduviidæ (Hemiptera-Heteroptera) from India and Ceylon'. *Verh. Naturfr. Ges. Basel* **68,** 122-31.

Miller N. C. E. 1957b, *Harmosticana garnhami* gen. nov., sp.n. (Lygæidæ: Hemiptera-Heteroptera) associated with a mammal in East Africa'. *Parasitology*, **46,** 206-208.

Miller N. C. E. 1958a, 'The Presence of a Stridulatory Organ in *Rhyticoris* (Hemiptera-Heteroptera-Coreidæ-Coreinæ)'. *Ent. month. Mag.* **94,** 238.

Miller N. C. E. 1958B, 'On the Reduviidæ of New Guinea and Adjacent Islands (Hemiptera-Heteroptera)'. *Nova Guinea* **9,** Pts. 1 and 2, 33-229.

Miller N. C. E. 1959, 'A New Subfamily, New Genera and New Species of Reduviidæ (Hemiptera-Heteroptera)., *Bull. Brit. Mus. Ent.* **8,** 2, 47-117.

Miller N. C. E. and Pagden H. T. 1931, 'Insect Remains in the Gut of a Cobra *Naia tripudians*'. *Nature, Lond.* 706.

Milliken F. B. and Wadley F. M. 1922, '*Geocoris pallens* Stål var. *decoratus* Uhl. a predaceous enemy of the false cinch bug'. *Bull. Brooklyn ent. Soc.* **17,** 143-6.

Mitis H. von 1935, 'Zur biologie der Corixiden, Stridulation'. *Z. Morph. Ökol. Tiere*, **30**, 479-95.
Miyamoto S. 1953, 'Biology of *Helotrephes formosanus* Esaki and Miyamoto with descriptions of larval stages'. *Sieboldia* **1**, 1, 10.
Mjöberg E. 1906, 'Über *Systellonotus triguttatus* L. und sein Verhaltnis zu *Lasius niger*'. *Zeitsch. f. wiss. Insektenbiol.* **2**, 107-109.
Mjöberg E. 1914, 'Preliminary description of a new representative of the family Termitocoridæ'. *Tijdschr. Ent.* **35**, 98-9.
Monod Theo. and Carayon J. 1958, 'Observations sur les *Copium* (Hemiptera-Tingidæ) et leur action cécidogène sur les fleurs de *Teucrium* (Labiées)'. *Arch. Zool. Exper. Gen.* **95**, 1-31.
Morrison J. 1932, 'Three apparently new species of Termitaphis'. *Zoologica* **3**, 20, 403-8.
Muir F. 1907, 'Notes on the stridulatory organ and stink-glands of *Tessaratoma papillosa* Thunberg'. *Tran. r. ent. Soc. Lond.*, Pt. 2.
Myers J. G. 1924, 'On the Systematic Position of the Family Termitaphidæ (Hemiptera-Heteroptera) with a description of a new genus and species from Panama'. *Psyche, Camb. Mass.* **31**, 6: 259-78.
Myers J. G. 1926a, 'Biological Notes on New Zealand Heteroptera'. *Trans. N.Z. Inst.* **56**, 449-54.
Myers J. G. 1926b, 'Heteroptera in Ocean Drift'. *Psyche* **33**, 110-5.
Myers J. G. 1929, 'Facultative Blood-sucking in Phytophagous Hemiptera'. *Parasitology* **21**, 472-80.
Myers J. G. 1932, 'Observations on the Family Termitaphidæ (Hemiptera-Heteroptera with the description of a new species from Jamaica'. *Ann. Mag. nat. Hist.* (10) 9, 366-72.

Nuorteva Pakka 1956, 'Studies on the Comparative Anatomy of the salivary glands in four families of Heteroptera'. *Ann. Ent. Fennica* **22**, 3, 45-54.
Nuorteva Pakka 1958, 'Die Rölle der Speichelsekrete in Wechselverhaltnis zwischen Tier und Nahrungspflanze bei Homopteren und Heteropteren'. *Ent. exp. appl.* **1**, 41-49.

Odhiambo T. R. 1957, 'The Bionomics of *Oxycarenus* species (Hemiptera-Lygæidæ) and their status as cotton pests in Uganda'. *Journ. Ent. Soc. S. Africa* **20**, 235-49.
Odhiambo T. R. 1958, 'Some Observations on the Natural History of *Acanthaspis petax* Stål (Hemiptera-Reduviidæ) living in termite mounds in Uganda'. *Proc. r. ent. Soc. Lond.* (A), **33**, 167-75.
Odhiambo T. R. 1958, 'The Camouflaging Habits of *Acanthaspis petax* Stål (Hem. Reduviidæ) in Uganda'. *Ent. month. mag.* **94**, 47.
Ohm, D. 1956, 'Beitrage zur Biologie der Wasserwanzen *Aphelocheirus æstivalis*'. *F. zool. Beitr. N. F.* **2**, 359-68.
Olivier E. 1899, 'Faune de L'Allier: Les Hemipteres Heteropteres'. *Rev. sci. Bourb.* **12**, 261.
Oshanin B. 1912, *Katalog dae Paläarktischen Hemiptera.*

Pagden H. T. 1928, '*Leptoglossus membranaceus* F. a pest of Cucurbitaceæ' *Malayan Agric. Journ.* **16**, 387-403.
Parshley H. M. 1917, 'Insects in Ocean Drift'. *Canad. Ent.* **49**, 45-8.
Pendergrast J. G. 1958, 'An egg-burster in Rhopalimorpha Dallas (Hem. Acanthosomidæ)'. *Ent. mon. Mag.* **94**, 72.
Pendergrast J. G. 1963, 'Observations on the biology and immature stages of *Antestia orbona* Kirk'. *N.Z. Ent.* **3**, 19-25.
Perez C. 1904, 'Sur les Phlœa, Hémiptères mimétiques de lichens'. *C.R. Soc. Biol. Paris*, **56** (1), 429-30.
Poisson R. 1924, 'Contribution à l'étude des Hémiptères Aquatiques'. *Bull. biol. France et Belgique*, **58**, 49-305.
Poisson R. 1930a, 'Sur un *Herpetomonas* parasite en Normandie de *Spilostethus* (*Lygæus*) *saxatilis* (Scop.) (Hémiptères Lygæoideæ). Apropos des Phytoflagelloses'. *C.R. Soc. Biol., Paris*, **103**, 1057-61.

Poisson R. 1930b, '*Herpetomonas tortum* n.sp. parasite intestinal des *Camptopus lateralis* (Germ.) (Hemiptera, Coreidæ, Alydaria) des environs de Banyuls. Rôle possible de cet insecte comme agent transmitteur de Phytoflagellose'. *C.R. Soc. Biol., Paris*, **103,** 1061-64.

Poisson R. 1935, 'Les Hémiptères Aquatiques (Sandaliorrhyncha) de la faune francaise'. *Arch. Zool. exp. gen.* **70,** 2, 480.

Poisson R. 1951, 'Traite de Zoologie, Héteroptères'. Tome **10,** fasc. **11,** 1675-1803.

Poppius B. 1909, 'Beitrage zur Kenntnis des Anthocoriden'. *Acta Soc. Sci. fenn.* 37, (9), 1-43.

Putshkov V. G. and Putshkova L. V. 1956, 'Eggs and larvæ of the Heteropera,. Pests of Agricultural Plants'. *Trud. Vsesouz. Ent. Obshch.* **45,** 218-342.

Putshkova L. V. 1955, 'Eggs of Hemiptera-Heteroptera I. Coreidæ'. *Ent. Obozr* **34,** 48-55.

Putshkova L. V. 1956, 'The Eggs of Hemiptera-Heteroptera II, Lygæidæ'. *Ent. Obozr.* **35,** 262-84.

Putshkova L. V. 1957, 'Eggs of Hemiptera Heteroptera III, Coreidæ (Supplement), IV (Macrocephalidæ), *Ent. Obozr.* **36,** 44-58.

Putshkova L. V. 1959, 'The eggs of the true bugs (Hemiptera Heteroptera) V, Pentatomidæ'. *Ent. Obozr.* **38,** 634-648.

Putschkova L. V. 1966, 'The morphology and biology of the eggs of the terrestrial bugs (Hemiptera)'. *Horæ Soc. ent. Un. Sovet.* **51,** 75-132.

Rabaud E. 1916, 'Le phénomène de la simulation de la mort'. *C.R. Soc. Biol.* **79,** 74.

Readio P. A. 1926, 'Studies on the Eggs of some Reduviidæ (Heteroptera)'. *Kans. Univ. Sci. Bull.* **16,** 4, 157-79.

Readio P. A. 1927a, 'Studies on the Biology of the Reduviidæ of America north of Mexico'. *Kans. Univ. Sci. Bull.* **17,** 1, 5-291.

Readio P. A. 1927b, 'Biological Notes on *Phymata erosa* sub-sp. *fasciata*'. *Bull. Brooklyn ent. Soc.* **2,** 256-62.

Readio P. A. 1931, 'Dormancy in *Reduvius personatus* L.' *Ann. ent. Soc. Amer.* **24,** 19-39.

Reuter O. M. 1885, 'Monographia Anthocoridarum orbis terrestris'. *Acta Soc. Sci. fenn.* **14,** 555-78.

Reuter O. M. 1909, 'Quelques mots sur les Phyllomorphes (Hem. Coreidæ)'. *Bull. Soc. ent. Fr.* 264-68.

Reuter O. M. 1912, 'Bermeikungen uber mein neues Heteropterensystem'. *Öfvers. Finska. Vetensk.-Soc. Förh.* (A), 54, (6), 1.

Reuter O. M. and Poppius B. 1909. 'Monographia Nabidarum orbis terrestris'. *Acta Soc. Sci. fenn.* **37,** No. 2, 1-62.

Roepke W. 1932, 'Uber Harzwanzen von Sumatra und Java'. *Misc. zool. Sumatra.* **68.**

Roonwal M. L. 1952, 'The Lantana Bug, *Teleonemis scrupulosa* Stål (*lantana* Distant) (Hemiptera-Tingidæ) with a description of its eggs, nymphs and adult'. *Journ. zool. Soc. India*, **4,** (1), 1-16.

Rosenkranz W. 1939, 'Die Symbiose der Pentatomiden (Hemiptera-Heteroptera'. *Zeitschr. Morph. Ökolog. der Tiere*, **36,** 2, 279-309.

Rubsaamen E. H. 1895, 'Über russische Zooceciden und deren Erzeuger'. *Bull. Soc. Imp. Moscova*, Nov., ser. 9, 396-488.

Ryckman, Raymond E. 1951, 'Recent Observations of Cannibalism in *Triatoma* (Hemiptera; Reduviidæ)'. *The Journal of Parasitology*, **37,** 5, 433-4.

Ryckman Raymond E. 1954, 'Lizards: A Laboratory Host for Triatominæ and *Trypanosoma cruzi* Chagas (Hemiptera, Reduviidæ), (Protomonadida: Trypanosomidæ)'. *Trans. Amer. Microsc. Soc.* LXXIII, 2, 215-8.

Ryckman Raymond E. 1958, 'Description and Biology of *Hesperocimex sonorensis*, New Species, An ectoparasite of the Purple Martin (Hemiptera Cimicidæ)'. *Ann. ent. Soc. Amer.*, **51,** 1, 33-47.

Ryckman Raymond E. and Ryckman Albert E. 1961, 'Baja California Triatominæ (Hemiptera: Reduviidæ), and their Hosts (Rodentia: Cricetidæ)'. *Ann. ent. Soc. Amer.* **54,** 1, 142-3.

Ryckman R. E. 1962, 'Biosystematics and hosts of the *Triatoma protracta* complex in North America (Hemiptera Reduviidæ), (Rodentia, Cricetidæ). *Univ. Cal. Publ. Ent.* **27,** 93-189.

Sailer R. I. 1945, 'The Bite of the Lacebug *Corythuca cydoniæ* (Fitch)'. *Journ. Kansas ent. Soc.* **18,** 81-2.

Sands W. H. 1957, 'The Immature Stages of some British Anthocoridæ (Hemiptera)'. *Trans. R. ent. Soc. Lond.* **109** (10), 295-310.

Saunders F. 1892, *The Hemiptera-Heteroptera of the British Islands.*

Saunders F. 1893, 'Antennal stridulation by *Centrocoris spiniger* F. *Ent. Month. Mag.* **29,** 99.

Saunders E. 1909, '*Myrmecoris gracilis* in the nest of *Formica rufa*'. *Ent. Month. Mag.* **39,** 269.

Schaefer Carl 1962, 'The Occurrence of an Axillary Spur in the Heteroptera and its Function in the Coreinæ'. *Ann. ent. Soc. Amer.* **55,** 6, 675-8.

Schaefer C. W. 1965, 'The Morphology and Higher Classification of the Coreoidea (Hemiptera, Heteroptera). Pt. III. The Families Rhopalidæ, Alydidæ and Coreidæ'. *Misc. Publ. ent. Soc. Amer.* **5,** No. 1, 1-76.

Schaefer C. W. 1966, 'The Morphology and Higher Systematics of the Idiostolinæ (Hemiptera, Lygæidæ). *Ann. ent. Soc. Amer.* **59,** 602-13.

Schneider H. 1928, 'Über die Zirporgan von *Piesma quadrata*'. *Fieb. Zool. Anz.* **75,** 329-330.

Schneider G. 1940, 'Beitrage zur Kenntnis der Symbiontischen Einrichtungen der Heteropteren'. *Zeitschr. Morp. Ökolog. der Tiere.* **36,** 4, 595-644.

Schorr H. 1957, 'Zur Verhaltens biologie und Symbiose von *Brachypelta aterrima* Forst. (Cydnidæ, Heteroptera)'. *Z. Morph. Ökol. Tiere.* **15,** 561-602.

Schouteden H. 1931, *Ann. Mus. Congo belg. Zool.* (3), 148.

Scudder G. G. E. 1962, 'Results of the Royal Society Expedition to Southern Chile, 1958-1959. Lygæidæ (Hemiptera), with the Description of a New Subfamily'. *Canad. Entom.* **94,** 1064-1075.

Scudder G. G. E. 1963, 'Pamphantinæ, Bledionotinæ and the Genus *Cattarus* Stål (Hemiptera; Lygæidæ)'. *Opus Ent.* XXVIII, 81-9.

Seidenstucker Gustav 1960, 'Heteropteren aus Iran 1956. *Thaumastella aradoides* Horv., eine Lygæid ohne ovipositor'. *Stuttgarter Beitr. z. Naturkunde.* **38,** 1-4.

Severin H. P. and Severin H. C. 1910, '*Notonecta undulata* Say preying on the egg of *Belostoma* (*Zaitha auctt.*) *flumineum* Say'. *Canad. Ent.* 240.

Shun-ichi-Nakao 1954, 'Biological and Ecological Studies on *Agriosphodrus dohrni* Signoret (Reduviidæ, Hemiptera). *Sci. Bull. I, II, Faculty Agric. Kyushu Univ.* **14,** 319-36.

Sikes E. E. and Wigglesworth V. B. 1931, 'The hatching of insects from the egg and the appearance of air in the tracheal system'. *Quart. Journ. Micr. Soc. Lond.* **74,** 165-92.

Silvestri F. 1911, 'Sulla posizione sistematica del genere *Termitaphis* Wasm. (Hemiptera) con descrizioni di due specie nuove'. *Boll. Lab. Zool. Portici,* **5,** 231-6.

Silvestri F. 1921, 'A New Species of *Termitaphis* (Hemiptera-Heteroptera) from India'. *Rec. Indian Mus.* **22,** 71-4.

Slater Fl. W. 1899, 'The egg-carrying habit of *Zaitha*'. *Amer. Nat.* **35,** 931-3.

Slater James A. 1951, 'The Immature Stages of American Pachygronthinæ (Hemiptera-Lygæidæ), *Iowa Acad. Sci.* **58,** 553-61.

Slater James A. 1955, 'The Macropterous form of *Lampracanthia crassicornis* (Uhler), (Hemipt. Saldidæ)'. *Journ. Kansas ent. Soc.* **28,** 3, 108.

Slater J. A. 1964, 'Hemiptera (Heteroptera) Lygæidæ. South African Animal Life; Results of the Lund Univ. Exped. in 1950-1951.

Slater James A. and Drake Carl J. 1956. 'The Systematic Position of the Family Thaumastocoridæ (Hemiptera-Heteroptera)'. *Proc. Tenth Int. Cong. Ent.* **1,** (1956), (1958).

Southall J. 1730, *A Treatise of Bugges.*

Southwood T. R. E. 1949, 'Some Notes on the Early Stages and Biology of *Sehirus bicolor* (L.), (Hem. Nabidæ)'. *Ent. mon. Mag.* **85,** 39-41

Southwood T. R. E. 1953, 'Interspecific Copulation between *Nabis ferus* (L.) and *N. rugosus* (L.) (Hem. Nabidæ)'. *Ent. mon. Mag.* **89, 294.**

Southwood T. R. E. 1955, 'The Egg and First Instar Larva of *Empicoris vagabundus* (L.), Hem. Reduviidæ'. *Ent. mon. Mag.* **91,** 96-7.

Southwood T. R. E. 1956, 'The Strutiure of the Eggs of the Terrestrial Heteroptera and its Relationship to the Classification of the Group'. *Trans. Roy. ent. Soc. Lond.* **108,** 163-221.

Southwood T. R. E. and Hine D. J. 1950, 'Further Notes on the Biology of *Sehirus bicolor* (L.) (Hem. Cydnidæ)'. *Ent. mon. Mag.* **86, 299-300.**

Southwood T. R. E. and Leston D. 1959, 'Land and Water Bugs of the British Isles,' *London*, pp. xii-436.

Southwood T. R. E. and Scudder G. G. E. 1956,' The Bionomics and Immature Stages of the Thistle Lace Bugs (*Tingis ampliata* H.S. and *T. cardui* L. (Hem. Tingidæ)'. *Trans. Soc. Brit. Ent.* **12,** (3), 93-112.

Speiser P. 1904, 'Die Hemipterengattung *Polyctenes* Gigl. und ihre stellung in System'. *Zool. Jahrb. suppl.* **7,** 373-380.

Spinola M. 1837, *Essai sur les Genres d'Insectes* 39-40.

Spooner C. S. 1938, 'The Phylogeny of the Hemiptera based on a study of the head capsule'. *Univ. Ill. Bull.* **35,** No. 70; 1-102.

Sprague I. B. 1956, 'The Biology and Morphology of *Hydrometra martini* Kirkaldy, *Univ. Kans. Sci. Bull.* **38,** 579-693.

Stusak J. M. 1957, 'Beitrag zur Kenntnis der Eier der Tingiden (Hemiptera-Heteroptera, Tingidæ)'. *Acta Soc. ent. Csl.* **55,** 361-71.

Stusak J. M. 1962, 'Immature Stages of *Elasmotrophis testacea* (H.-S.) and Notes on the Bionomics of the Species (Het. Tingidæ)'. *Acta Soc. ent. Csl.* **59,** 19-27.

Steer W. 1929, 'The Eggs of Some Hemiptera-Heteroptera'. *Ent. mon. Mag.* **65,** 34-8.

Stroyan H. L. G. 1954, 'Notes on the Early Stages of *Rhopalus parumpunctatus* Schill. (Hemiptera-Coreidæ)'. *Proc. Roy. ent. Soc. Lond.* (A), **29,** 32-8.

Stys P. 1961, 'The Stridulatory Mechanism in *Centrocoris spiniger* (F.) and some other Coreidæ (Heteroptera). *Acta ent. Mus. Nat. Prague* XXXIV, **592,** 427-31.

Stys P. 1964, 'On the Morphology and Taxonomy of Agriopocorinæ (Heteroptera, Coreidæ)'. *Cas. Cesk. Spol. Ent.* (*Acta Soc. ent. Cech.*), **61,** 25-38.

Stys P. 1964, 'Morphology and Relationship of the family Hyocephalidæ. *Act. Zool. Acad. Sci. Hung. X*, fasc. 1-2, 229-62.

Stys P. 1966, 'Revision of the Genus *Dayakiella* Horv., and Notes on its Systematical Position (Heteroptera, Colobathristidæ). *Acta ent. bohemoslov.* **63,** 27-39.

Stys P. 1967, 'Medocostidæ. A New Family of Cimicimorphan Heteroptera based on a new genus and two new species from Tropical Africa I. Descriptive Part'. *Acta ent. bohemoslov.* **64,** 439-65.

Sweet Merrill H. 1960, 'The Seedbugs; A Contribution to the Feeding Habits of Lygæidæ (Hemiptera-Heteroptera)'. *Ann. ent. Soc. Amer.* **53,** 3, 317-21.

Sweet M. H. 1964, 'The biology and ecology of the Rhyparochrominæ of New England (Heteroptera; Lygæidæ)', Part I. *Ent. Amer.* **43,** 1-124.

Szent-Ivany J. J. H. and Catley A. 1960, 'Notes on the Distribution and Economic Importance of the Papuan Tip-wilt Bug *Amblypelta lutescens papuensis* Brown (Heteroptera-Coreidæ)'. *Papua and New Guinea Agricultural J.* **13,** 2, 59-65.

Takahashi R. 1921, 'Observations on the Ochteridæ'. *Trans. nat. Hist. Soc. Formosa* **11,** 55, 119-25.

Tamanini L. 1956, 'Osservationi biologiche e morphologiche sugli *Aradus betulinus* Fall., *A. corticalis* L., *A. pictus* Bar. (Hem. Het. Aradidæ)'. *Stud. Trent. Sci. Nat.* **33,** 3-53.

Taylor T. H. C. 1945, 'Recent Investigations of *Antestia* species in Uganda, *E. Afric. Agric. J.* **10,** 4, 223-33.

Theobald F. 1895, 'Notes on the needle-nosed hop-bugs'. *J. S.-E. Agric. Coll., Wye*, 7 pp.

Thomas D. C. 1954, 'Notes on the Biology of some Hemiptera-Heteroptera'. *Entomologist* **87,** 25-30.

Thontadarya T. S. and Basavanna C. G. P. 1959, 'Mode of egg-laying in the Tingidæ (Hem.)'. *Nature* **184,** 289-90.

Thorpe W. H. and Crisp J. D. 1947, 'Studies on plastron respiration I. The Biology of *Aphelocheirus* (Hemiptera-Aphelocheiridæ, Naucoridæ) and the mechanism of plastron retention'. *Journ. Exper. Biol.* **24,** 227-69.

Thorpe W. H. 1950, 'Plastron respiration in aquatic insects'. *Biol. Reviews* **25,** 344-90.

Tischler W. 1960, 'Studien zur Bionomie und Okologie der Schmalwanze *Ischnodemus subuleti* Fall. (Hem. Lygæidæ) *Z. Wiss. Zool.* **163,** 168-209.

Tonapi G. T. 1959, 'A note on the eggs of *Gerris fluviorum* F. with a brief description of the neanide. (Hem. Gerridæ)'. *Ent. mon. Mag.* **95,** 29-31.

Usinger R. L. 1934, 'Bloodsucking among phytophagous Hemiptera'. *Canad. Ent.* **66,** 97-100.

Usinger R. L. 1932, 'Miscellaneous Studies in the Henicocephalidæ (Hemiptera)'. *Pan. Pacif. Ent.* **8,** 4, 145-56.

Usinger R. L. 1938, 'Biological Notes on the pelagic water striders (*Halobates*) of the Hawaiian Islands with description of a new species from Waikiki (Gerridæ-Hemiptera)'. *Proc. Hawaii ent. Soc.* **10,** 77-84.

Usinger R. L. 1939, 'Distribution and Host Relationships of *Cyrtorhinus* (Hemiptera, Miridæ)'. *Proc. Harv. ent. Soc.* **10,** 271-3.

Usinger R. L. 1941, 'Three New Genera of Apterous Aradidæ. *Pan. Pacif. Ent.* **17,** 169-81.

Usinger R. L. 1942a, 'The genus *Nysius* and its allies in the Hawaiian Islands (Hemiptera, Lygæidæ, Orsillini)'. *Bull. Bishop Mus. Honolulu* **173,** 1-167.

Usinger R. L. 1942b, 'Revision of the Termitaphidæ (Hemiptera)'. *Pan. Pacif. Ent.* **18,** 155-9.

Usinger R. L. 1943, 'A Revised Classification of the Reduvioidea with a New Subfamily from South America'. *Ann. ent. Soc. Amer.* **36,** 4, 602-18.

Usinger R. L. 1944, 'The Triatominæ of North and Central America and the West Indies and their Public Health Significance'. *Publ. Hlth. Bull. Wash.* 288.

Usinger R. L. 1946a, 'Insects of Guam 2. Hemiptera-Heteroptera'. *Bull. Bishop Mus. Honolulu* **189,** 11-103.

Usinger R. L. 1946b, 'Biology and Control of Ash Plant Bugs in California'. *J. Econ. Ent.* **38,** (5), 585-91.

Usinger R. L. 1946c, 'Notes and descriptions of *Ambrysus* Stål with an account of the life history of *Ambrysus mormon* Montd. (Hemiptera, Naucoridæ)'. *Kansas Univ. Sci. Bull.* **31,** 185-210.

Usinger R. L. 1947a, Biology and Control of the Ash Lace Bug *Leptophya minor*'. *J. Econ. Ent.* **30,** (3), 286-9.

Usinger R. L. 147b, 'Native Hosts of the Mexican Chicken Bug *Hæmatosiphon inodora* (Duges) (Hemiptera Cimicidæ)'. *Pan Pacif. Ent.* **23,** 3, 140.

Usinger R. L. 1950, 'The Origin and Distribution of Apterous Aradidæ. *8th Int. Congr. Ent.* 1-6.

Usinger R. L. 1952, 'A New Species of *Carayonia* from Ceylon'. *Entomologist* **85,** 212.

Usinger R. L. 1954, 'A New Genus of Aradidæ from the Belgian Congo with Notes on the Stridulatory Mechanisms of the Family'. *Ann. Mus. Congo Belge, Zool.* I, Misc. Zoologica H. Schouteden 540-3.

Usinger R. L. 1958, ' "Harzwanzen" or "Resin Bugs" in Thailand'. *Pan. Pacif. Ent.* **43,** 52.

Usinger R. L. 1959, 'New Species of Cimicidæ (Hemiptera)'. *Entomologist* **92,** 218-22.

Usinger R. L. and Aslock P. D. 1959, 'Revision of the Metrargini (Hemiptera Lygæidæ)'. *Proc. Hawaii ent. Soc.* **17**, 93-116.
Usinger R. L. and Herring J. L. 1956, 'Notes on Marine Water Striders of the Hawaiian Islands (Hemiptera-Gerridæ)'. *Proc. Hawaiian ent. Soc.* **16**, 281-3.
Usinger R. L. and Matsuda 1959, Clas. Aradidæ, *Brit. Mus.*
Usinger Robert L., Wygodzinsky P. 1964, 'Description of New Species of *Mendanocoris* Miller with Notes on the Systematic Position of the Genus. (Reduviidæ, Hemiptera, Insecta)'. *Amer. Mus. Novit.* **2204**, 1-13.
Usinger R. L. *et al.* 1966, 'Monograph of Cimicidæ (Hemiptera-Heteroptera) *Thomas Say Foundation*, Vol. VII.

Vecht J. van de 1953, 'Het Lantana-Wantsje in Indonesie (*Teleonemia scrupulosa* Stål, Fam. Tingidæ)'. *Tijdschr. over Plantenziekten*, **59**, 170-3.
Verhoeff G. 1893, 'Vergleichende Untersuchungen über die Abdominal-Segments der Weiblichen Hemiptera-Heteroptera und Homoptera'. *Verh. naturh. ver. Rheinl. Westf. auch Diss Bonn.* 307-74.
Villiers A. 1945, 'Un Nouvel Holoptilide (Hém.) du Sud-Ouest de l'Afrique'. *Bull. Soc. ent. Fr.* No. 7, 106-7.
Villiers, A. 1951, 'Sur deux Réduvidés Saicinæ du Cameroun récoltés par J. Carayon'. *Bull. Mus. Paris* (2), 3, 279.
Villiers A. 1952, *Hémiptères de l'Afrique noire* (*Punaises et Cigales*). *Initiations africains.* Institut. franç. d'Afrique noire, Dakar.
Villiers A. 1958a, 'Insectes Hémiptères Enicocephalidæ.' *Faune de Madagascar.* **7**, 5-78.
Villiers A. 1958b, 'Hémiptères Réduvidés récoltés en Angola (3eme note). *Publ. Cult. Comp. Diam. Angola*, **14**, 44.
Villiers A. 1962, 'Henicocephalidæ (Hemiptera-Heteroptera)'. *Parc National de la Garamba Mission H. de Saeger*, Fasc. 32, (i), 3-26.
Vodjani S. 1964, 'Contribution à l'étude des punaises de Cereales et en particulier d'*Eurygaster integriceps* Put. (Hemiptera, Pentatomidæ, Scutellerinæ)'. *Ann. Epiphyt.* **1954**, 105-60.

Walton G. A. 1936, 'Oviposition in the British Species of *Notonecta* (Hemiptera)'. *Trans. Soc. Brit. Ent.* **3**, 49-57.
Walton G. A. 1962, 'The egg of *Agraptocorixa gestroi* Kirkaldy (Hemiptera; Heteroptera; Corixidæ)'. *Proc. Roy. ent. Soc. Lond.* (A), **37**, 104-6.
Wasman E. 1911, 'Die Ameisen und ihre Gäste'. *Ist. Int. Ent. Congr.* 209-304.
Waterhouse C. O. 1897, 'On the affinity of the genus *Polyctenes* Giglioli with a description of new species'. *Trans. ent. Soc. Lond.* 309-312.
Weber H. 1930, *Biologie der Hemipteren.*
Wefelscheid H. 1912, 'Uber die Biologie und Anatomie von *Plea minutissima* Leach'. *Zool. Jahrb.* **32**, 389-474.
Wendt A. 1939, 'Beitrag zur Kenntnis der Verbreitung und Lebensweise der Schwalbenwanze (*Oeciacus hirundinis* Jen.) in Mecklenburg'. *Archiv. d. Ver. d. Freunde der Naturgesch. in Mecklenburg. neue Folge* **14**, 71-94.
Wesenberg-Lund C. 1943, *Biologie der Susswasserinsekten, Kopenhagen.*
Westwood J. O. 1871, '*Corixa* destructive to the ova of fishes in India'. *Proc. ent. Soc. Lond.* III-IV.
Whitfield F. G. S. 1933, 'The Sudan Millet bug *Agnoscelis versicolor* F. *Bull. ent. Res.* **20**, 209-24.
Wigglesworth V. B. 1936, 'Symbiotic Bacteria in a blood-sucking insect *Rhodnius prolixus* Stål (Hemiptera-Triatominæ)'. *Parasitology* **28**, 2, 284-9.
Wigglesworth V. B. 1938, 'Climbing Organs in Insects'. *Nature, Lond.* 974.
Wigglesworth V. B. 1939, *Principles of Insect Physiology.*
Wigglesworth V. B. 1954, *The Physiology of Insect Metamorphosis.*
Wigglesworth V. B. and Beament J. W. L. 1950, 'The Respiratory Mechanism of Some Insect Eggs'. *Quart. J. micr. Sc.*, **91**, 4, 429-52.

Wille J. 1929, 'Die Rübenblattwanze *Piesma quadrata* Fieb. (Piesmatidæ)'. *Monog. Pfl. Sch.* **2,** 116 pp.

Wiley G. O. 1922, 'Life History Notes on Two Species of Saldidæ (Hemiptera) found in Kansas'. *Kansas Univ. Sc., Bull.* **14,** 299-310.

Woodroffe G. E. 1963, 'Biological notes on British Hemipetra-Heteroptera captured during 1962'. *Ent. mon. Mag.* **86,** 82-4.

Woodward T. E. 1952, 'Studies on the reproductive cycle of three species of British Heteroptera with special reference to overwintering stages'. *Trans. Roy. ent. Soc. Lond.* **103,** 171-218.

Woodward T. E. 1968, 'A new subfamily of Lygæidæ (Hemiptera-Heteroptera) from Australia'. *Proc. Roy. ent. Soc. Lond.* (B), **37,** (9-10), 125-32.

Wray D. L. and Brimley C. S. 1943, 'The Insect Inquilines and Victims of Pitcher Plants in N. Carolina'. *Ann. ent. Soc. Amer.* **36,** 128-37.

Wygodzinsky P. 1944, 'Notas sobre a biologia e o desinvolvimento do *Macrocephulus notatus* Westw. (Phymatidæ, Reduvioidea, Hemiptera)" *Rev. Ent. Rio de J.* **15,** 139-143.

Wygodzinsky P. 1944, 'Contribuçao ao Conhecimento do Genero "Elasmodema" Stål 1860 (Elasmodemidæ, Reduvioidea, Hemiptera)'. *Rev. bras. Biol.* **4,** (2), 193-213.

Wygodzinsky P. 1946a, 'Sobre duas novas especies de Emesinæ do Brasil com Notas sobre "*Stenolæmoides arizonensis* (Banks)", (Reduviidæ, Hemiptera)'. *Rev. bras. Biol.* **6,** (4), 509-19.

Wygodzinsky P. 1946b, 'Contribution towards the Knowledge of the Isoderminæ, Aradidæ-Hemiptera)'. *Rev. Ent. Rio de J.* **17,** 261-73.

Wygodzinsky P. 1947a, '*Trichotonannus setulosus* Reut. in nest of *Protermes minutus* Grassi'. *Rev. franç. ent.* **14,** 11, 120.

Wygodzinsky P. 1947b, 'Contribuʖao ao Conhecimento do Genero *Heniartes* Spinola (Apiomerinæ, Reduviidæ, Hemiptera)'. *Arch. Mus. nac. Rio de J.* **41**.

Wygodzinsky P. 1948, 'On Some Reduviidæ belonging to the Naturhistorisches Museum of Vienna (Hemiptera)'. *Rev. bras. Biol.* **8,** (2), 209-24.

Wygodzinsky P. 1950, 'Schizopterinæ from Angola (Cryptostemmatidæ Hemiptera)'. *Publ. cult. Cia. Diam. Angola* **7,** 9-47.

Wygodzinsky P. and Usinger R. L. 1963, 'Classification of the Holoptilinæ and Description of the first representative from the New World. (Hemiptera; Reduviidæ)'. *Proc. Roy. ent. Soc. Lond.* (B), **32,** 3-4, 49-52.

Yamada Y. 1914, 'On *Urostylus westwoodii* Scott'. *Insect World* **18,** 138-142.

Yamada Y. 1915, 'On *Urostylus striicornis* Scott'. *Insect World* **19,** 313-6.

Yang We-I 1936, 'The Outbreak of the *Urochela distincta* Distant in Lushan'. *Bull. Fan. Inst. Biol. Peking* **8,** No. 2, 57-61.

York Geo. T. 1944, 'Food Studies of *Geocoris* spp. Predators of the Beet leafhopper'. *J. Econ. Ent.,* **37,** 1, 25.

Yuasa H. 1929, 'An ecological note on *Speovelia maritima* Esaki'. *Ann. Mag. nat. Hist.* (10) **4,** 346-9.

Yuksel M. 1958, 'Biology, ecology and control of Senn Pest, *Eurygaster integriceps* in Turkey (1955/6)'. *Hofchenbr. Wiss.* **1,** 25-36.

INDEX